Underground Houses

How to Build a Low-Cost Home

Robert L. Roy

A Drake Publication
Sterling Publishing Co., Inc. New York

Oak Tree Press Co., Ltd.
London & Sydney

To my mother with love.
She must be wondering what crazy thing
we're going to do next.

Second Printing, 1979
Copyright © 1979 by Robert L. Roy
Published by Sterling Publishing Co., Inc.
Two Park Avenue, New York, N.Y. 10016
Distributed in Australia by Oak Tree Press Co., Ltd.,
P.O. Box J34, Brickfield Hill, Sydney 2000, N.S.W.
Distributed in the United Kingdom by Ward Lock Ltd.
116 Baker Street, London W.1
Manufactured in the United States of America
All rights reserved
Library of Congress Catalog Card No.: 79-64505
Sterling/Drake ISBN 0-8069-8856-8 Trade   Oak Tree 7061-2641-6

CONTENTS

Publisher's Note

While *Underground Houses: How to Build a Low-Cost Home* is the author's step-by-step account of his construction of one specific dwelling, it is a practical guide for building any type of earth-sheltered home as it does demonstrate the principles and techniques that are basic to all. The dimensions, calculations and materials must, of course, be adapted to suit the site, land contours, plumbing, space requirements and energy needs of each individual project.

"I even tell lies myself. I hear myself telling people the main reason to build underground is energy conservation, and I talk a lot about all the money and natural resources you can save. Or that underground buildings are so quiet. Or that I build them to reduce the rainwater run-off invariably associated with impervious roofs. Or because all the sea and land surfaces of the earth were meant to be home to green plants and not to asphalt and concrete.

"All those reasons are true—and terribly important—but none is my main reason for going under. I'm always afraid to mention the real reason for fear it will seem frivolous and drive people away. So I wait until I know I'm among friends before I admit I do it primarily because it is so beautiful. *Please do not tell anyone this secret reason!*"

Malcolm Wells
CoEvolution Quarterly, Fall 1976

INTRODUCTION

Subterranean housing is nothing new. Even low-cost subterranean housing is nothing new. After all, what did it cost the first cave man to walk into his shelter? Perhaps he had to do battle with a den of snakes or a sleeping bear. But that may well have been a better deal than a thirty-year mortgage.

But we leave this talk of caves as quickly as possible. We may joke about returning to cave-dwelling times, and my wife Jaki and I even call our habitation "Log-End Cave," but the common conception of underground housing as dark, damp, and dismal is itself a throwback to prehistoric times. Modern terratecture (subterranean architecture) goes out of its way to assure that the opposite conditions prevail. For my part, I am willing to compare the light and nonclaustrophobic space of Log-End Cave to those of more conventional surface dwellings. And we built it without battling snakes, bears, or a thirty-year mortgage! (A complete cost analysis of Log-End Cave appears in Appendix A and demonstrates that a 910-square-foot energy-efficient dwelling can be built for $6,750.)

Man has lived underground at many times and places throughout history. In Cappadocia, a subdesertic region of Turkey, people have been living in underground towns and cities since before Christ. The settlements, some of which extend eight or ten stories below ground, are hewn out of the soft stone that prevails in the area. These subterranean settlements are a response to a hostile environment, and the micro-climate of the villages remains constant and comfortable despite harsh variations of heat and cold on the surface.

Folk architecture has always made intelligent use of available materials. When the materials are no more than the soft strata itself, limestone perhaps, it is only natural that caves and underground dwellings evolve. In Sicily, there are chambers cut into nearly perpendicular walls of the Anapo Valley. Originally, these chambers served as burial grounds for an adjacent prehistoric town, but in the Middle Ages they were refashioned into dwellings.[1]

"Underground" writer Richard F. Dempewolff says: "In Gaudix, Spain, 30,000 people live in apartments supplied with electricity, tile floors, and all the creature comforts—carved in soft rock cliffs. There are now even *luxury* caves in France's Loire Valley where caverns, left by quarriers who cut stone for the great chateaus, have been extended, paneled, carpeted, richly furnished, and sold to wealthy city dwellers who cherish their coolness in summer and natural warmth on winter weekends." [2]

In the Canary Islands, cave dwelling is found in the agricultural uplands. I visited these areas ten or twelve years ago and remember thinking how uncavelike the dwellings were. It was as if a house had been set into the hillside, with only one wall left exposed. There were even curtains on the windows. Those houses were not as brightly lit as it is possible to make them today, with such developments as the insulated skylight, but they were cheery and will no doubt still be much the same hundreds of years from now. One wonders how many generations of little troglodytes have returned to the cool of their homes after a hot day on the terraces.

These glimpses into other societies are interesting, to be sure, but the question arises, Why would anyone want to build underground in modern America? Well, when the evidence is weighed, I think the question should be, Why wouldn't they? Malcolm Wells, the well-known architect and an "Underground Man," offers several reasons for building below the surface. Subterranean living, he observes, offers silence; freedom from vibration; wildlife habitat and green, living land in place of rats and asphalt; weather moderation and temperature buffering; and oxygen and even food production in place of the blistering heat of a lifeless roof. He points out that an earth roof makes the proper use of rainwater that we normally waste, and it encourages percolation and a slow runoff instead of erosion and flash floods. [3]

One might lengthen Wells's list by pointing out that an underground dwelling offers protection from fires, tornadoes, earthquakes, and—for the pessimistic—bombs. John Barnard, designer of the well-known "Ecology House" in Marstons Mills on Cape Cod, cites the absence of the need for exterior maintenance among the advantages of subterranean housing. Roofs don't need to be replaced, and pipes don't freeze. Nor do walls need to be painted or gutters cleaned.

But there are two advantages of subterranean dwelling that, in my opinion, outweigh all the rest. They are (1) heating and cooling efficiency and (2) low cost of construction for the owner-builder. This book documents the efficiency and is intended to familiarize the reader with techniques for achieving the low-cost construction.

Note: Work on this book was interrupted in mid-June of 1978 by the Towards Tomorrow Fair at the University of Massachusetts, where I helped Jack Henstridge with his presentation of log-end construction techniques. At the fair, I met people from the American Underground-Space Association, who kindly traded a copy of a report prepared by the Underground Space Center for my previous book, *How to Build Log-End Houses* (Drake, 1977).

Here's their blurb: "*Earth Sheltered Housing Design* is approximately 300 pages long and represents the results of a one-year design study carried out with funding from the Legislative Commission on Minnesota Resources. Plans and sections of twenty earth sheltered houses are included, with interior and exterior photographs of the completed houses. The report is intended to assist people in the layout and design of earth sheltered houses and has an emphasis on the design of houses for Minnesota. The implications of design for energy use, structural needs, and other problems are discussed. The public policy issues cover code, legal, financial and insurance aspects of earth-sheltered housing. An annotated bibliography and other sources of information are included to assist further research."

All I can add is that I wish the report had been in print a year earlier. It is the most thorough discussion of the principles involved in subterranean house construction that I have seen. It is not a how-to book, but I do not hesitate to advise potential subterranean owner-builders to send for a copy. It is epsecially valuable for those planning to build in areas where the Uniform Building Code is in force. In our area it is not, and the reader may spot floor plan features of our house which run contrary to the UBC. It is not difficult to satisfy the code with subterranean house plans, but the cost of the house is likely to rise somewhat because of certain escape-route regulations. Occasionally, in this book, I have added notes or given another point of view based on information appearing in *Earth Sheltered Housing Design*.

Copies of the study may be obtained from:

American Underground-Space Association
Department of Civil and Mineral Engineering
University of Minnesota
Minneapolis, Minnesota 55455

Robert L. Roy

1. Design

There are two different approaches to subsurface dwelling: the bermed house and the chambered house. The bermed house involves building the structure at or close to original grade and "berming" the side walls of the building with earth as well as covering the roof with earth and sod. In the chambered house, the entire structure is below the original grade.

There are four primary terratectural types that can be employed with either the bermed or the chambered house. These are shown in Figure 1, which is taken from a paper published by Kenneth Labs in *Alternatives in Energy Conservation: The Use of Earth Covered Buildings.*[4]

The "true" underground, or vault, is internally similar to deep space because of its isolation. There is no direct contact with the outside world. This terratectural type is probably not the best for private dwellings, although it has been used successfully for industrial parks and shopping centers.

The atrium or courtyard style of terratecture is popular among exponents of underground living. John Barnard's Ecology House is a chambered dwelling incorporating an atrium for entry, light, and air; this central courtyard also becomes a kind of inside/outside living space.

The elevational design incorporates windows and walk-in doors. The open wall should face south for maximum use of light and solar energy. Our own Log-End Cave is of the elevational type, somewhere midway between the bermed and the chambered approach.

Side-wall penetration, according to Labs, provides light, air, access, view, and the potential for expansion. Malcolm Wells's underground architect's office (Fig. 2) in Cherry Hill, N.J., makes good use of these side-wall penetrations.

We decided that the elevational type was best suited to our land and to ourselves. It enabled us to utilize an ideal south-sloping knoll and afforded us a view into the forest. The atrium and side-wall penetration types are not well suited to providing the

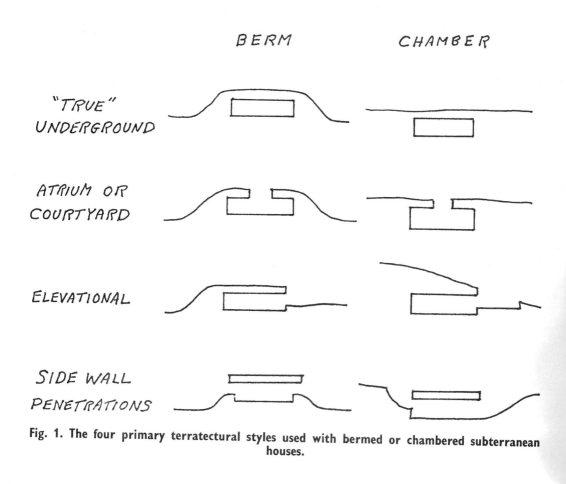

BERM   CHAMBER

"TRUE"
UNDERGROUND

ATRIUM OR
COURTYARD

ELEVATIONAL

SIDE WALL
PENETRATIONS

Fig. 1. The four primary terratectural styles used with bermed or chambered subterranean houses.

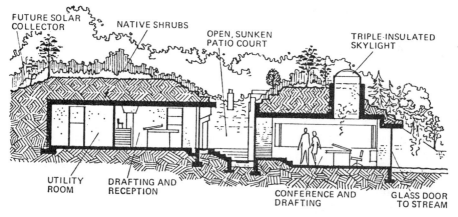

FUTURE SOLAR COLLECTOR   NATIVE SHRUBS   OPEN, SUNKEN PATIO COURT   TRIPLE-INSULATED SKYLIGHT

UTILITY ROOM   DRAFTING AND RECEPTION   CONFERENCE AND DRAFTING   GLASS DOOR TO STREAM

Fig. 2. The underground office of architect Malcolm Wells features a sunken patio and a skylight that admits sunlight and air to all rooms; a sliding door opens out on a flowing stream. (Reprinted with permission from "Your Next House Could Have a Grass Roof," Richard F. Dempewolff, Popular Mechanics © 1977 by The Hearst Corporation.)

external view that we feel contributes to the airy and open atmosphere at Log-End Cave. Although every owner-builder likes to design his own home, incorporating the features he finds important, we offer our own plans (Figs. 3 and 4) to illustrate certain construction techniques common to the subterranean idea. This house can be built on a gentle slope, as ours was, or on a flat site as a bermed structure.

Our plans integrate site considerations, the availability of materials, certain goals for efficiency in heating and cooling, and our desire for a floor plan that would be easy to live with. Our north-south dimensions were limited by the availability of three 30-foot 10-by-10-inch barn beams, which we wanted to use as the main load-supporting timbers. Our east-west dimension of 30 feet was limited by two structural considerations: our desire to build without pilasters (internal buttresses, discussed later), and considerations for the strength of the roof rafters. Neither of these limitations interfered with our plans to create a roomy two-bedroom home. Had longer beams been available, we could have gone to a 35-foot-square house, increasing our internal square footage from 910 to 1,090. I felt that at 40 feet on a side (1,444 square feet with 1-foot external walls) pilasters would have been necessary. A three-bedroom house of similar design could be made by converting the office or sauna to a third bedroom. Or a roomy 40-foot square house could be built with three bedrooms, the office, and the sauna. Such a house would be over 50 percent larger than our own, and I think it would be safe to assume that it would cost about 50 percent more to build and require almost 50 percent more fuel to heat. Building time would only be about 25 percent greater, although the need for pilasters could increase that estimate somewhat. There is no reason why the main support beams could not be in two pieces joined by a pegged mortise and tenon joint, but the rafters—4-by-8s at Log-End Cave —would then have to be 4-by-10s, which are extremely heavy and difficult to work with.

Our floor plan was carefully worked out to keep the peripheral rooms about 8 degrees lower in temperature than the open-plan living/kitchen/dining area. We prefer to sleep in cooler bedrooms and believe that they are healthier, and we do not see any need to keep the other peripheral rooms (sauna, bathroom, larder, and mudroom) at 70°F. With the heat source in the center, the slowdown of heat transfer through the internal walls maintains the desired difference in temperature and makes the best use of our fuel.

Our floor plan was also designed to make the intersection of internal walls and exposed ceiling both neat and easy to construct. This will be discussed later.

As stated earlier, Log-End Cave is an example of elevational terratecture. It is not a "true" underground house in that we utilize windows and skylights for natural lighting. In addition, the south side of our house is less than halfway underground, although it is mostly below the original grade of the land. Our single entrance is provided by a further sculpturing of the south-side excavation, so that there is actually a slight grade upwards to the house. This is important in preventing water seepage via the entrance. Water cannot collect outside the door. In fact, there are perforated tile drains as well as a slope to carry away a fast accumulation of surface water, such as might occur during a warm spring rain.

I am aware of the disadvantages of a single-entrance home, not the least of which is that this violates the safety reasons behind some building codes. Remember, though,

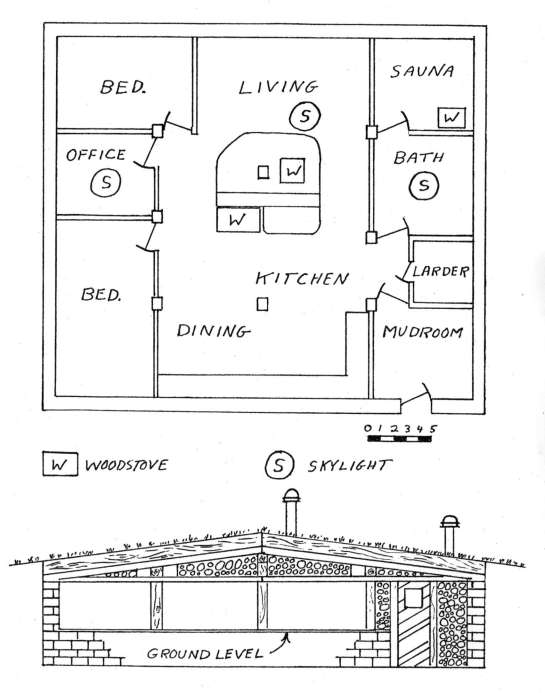

BED.

LIVING

Ⓢ

SAUNA

W

OFFICE

Ⓢ

W

W

BATH

Ⓢ

W

BED.

KITCHEN

□

DINING

LARDER

MUDROOM

0 1 2 3 4 5

W WOODSTOVE   Ⓢ SKYLIGHT

GROUND LEVEL

Fig. 3. The floor plan and south-wall elevational plan of Log-End Cave, a chambered structure built on a gentle slope.

11

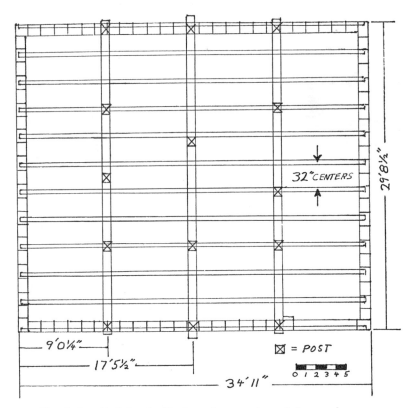

Fig. 4. Log-End Cave's block, rafter, post, and beam plan.

that building codes have been created almost exclusively for above-ground structures. We were not hampered by building codes in our area and felt that the inherent safety of our cave against the danger of fire made a second entrance unnecessary. Of course, such a second entrance would entail more time, work, and money, as well as provide an additional troublespot for possible leakage and heat loss. If necessary, a second entrance could be incorporated in the south wall of a house of similar design, either through the bedroom or by some change in the interior layout.

Another way to overcome the problem of a second entrance is to run a stairway upwards, perhaps where the office is in our plans, to a small above-ground building. This building could be round or rectilinear and could serve as a seasonal living area, guest room, or porch. Such a building would not have to be insulated if used in that fashion, as long as the door leading to the stairway was well insulated. Egress from the little building would lead onto the grass roof of the main dwelling. If the little building was tastefully designed—with a sod roof, for example—it could be an interesting design feature. I have no quarrel with purists who might argue that such a structure would be a blot on an otherwise undistrubed landscape; I tend to feel that way myself. But others may differ, and I submit the idea for whatever it's worth.

While building, I tried to keep in mind three main considerations of primary importance in subterranean home construction: strength (because of the extra heavy roof load), waterproofness, and livability. For underground housing, I define "livability"—not found in *Webster's Unabridged*—as enough natural light and ventilation to guarantee a nonclaustrophobic atmosphere.

2. Siting

We spent a lot of time choosing our general site and even more time when it came to actually placing our four flags on the ground as guides for the front-end-loader operator.

We live on a 44-acre homestead in the foothills of the Adirondacks. Most of the land is wooded, except for about two acres where we do our gardening and build the various structures that are a part of our chosen lifestyle. We'd already built one house, Log-End Cottage, in this two-acre compound. We decided to build Log-End Cave about 125 feet from the cottage, in order to take advantage of a common driveway, well, septic system, and windplant.

Readers who have seen my previous book, *How to Build Log-End Houses,* may wonder why we decided to leave the cottage in favor of the cave. Well, our family increased by 50 percent (a boy, Rohan), and our little cottage suddenly got a lot smaller. And the layout of the cottage was not conducive to homesteading activities. We always have several indoor projects going—canning, beer making, and the like— and they always seem to center at the kitchen table. At the same time, we had come to appreciate the tremendous heating and cooling advantages of underground living. It took seven full cords of firewood to keep us comfortable during the unusually cold winter of 1976–1977, which was not too bad, but we figured to cut this to two or three in an underground house of greater living area.

For the benefit of those who have been seriously thinking about building with log-ends, I would like to say that our moving out of the cottage had nothing whatsoever to do with any fault in that style of building. On the contrary, we are as enamored as ever of this beautiful and inexpensive building technique and have retained log-ends as an important part of the construction of the cave.

About 125 feet from the cottage, just beyond the windplant tower, is the top of a knoll, the highest point for at least a quarter mile in any direction. This knoll slopes gently in all directions but, luckily, its greatest slope is just west of south. This site, then, combined the advantage of height—good drainage away from the

Table 1. Statistical Abstract of the Building Site

*Contour	0°	−15°	−30°	−45°	−60°	−75°	−90°
1"	6' 1"	6' 8"	7' 7"	9' 4"	12' 2"	16' 5"	23' 7"**
6"	15'	15' 11"	18' 6"	21' 6"	24' 9"	29' 4"	33' 4"
12"	19' 4"	21'	22' 7"	27' 11"	32' 8"	34' 2"	36'
18"	24' 2"	26'	27' 11"	31' 10"	36'		
24"	27' 10"	30' 2"	33'	36' 1"			
30"	30' 4"	32' 4"	37'	41' 4"			
36"	35' 2"	38' 2"	40' 10"	46' 4"			
42"	40' 6"	42' 5"	47'				
48"	46' 7"	49' 2"	54'				

*Contour	+15°	+30°	+45°	+60°	+75°	+90°
1"	6'	6' 3"	6' 6"	7'	7' 8"	8'
6"	15' 7"	16' 2"	17' 2"	16' 7"	17'	21' 6"
12"	22' 6"	21' 7"	20' 11"	21' 10"	27' 2"	29'
18"	24' 4"	23'	22' 5"	26' 3"	31' 4"	35' 8"
24"	26' 6"	25' 10"	25' 11"	30' 1"	36'	40' 8"
30"	29' 9"	29' 1"	29' 4"	34'	38' 10"	45' 3"
36"	34' 8"	32' 8"	33' 7"	37' 5"	42' 4"	49' 7"
42"	38' 2"	36' 8"	36' 3"	39' 5"	46' 4"	
48"	43' 1"	40' 10"	40' 1"	41' 9"		

* Contours in inches below reference point.
** Distance from reference point to certain contours along primary rays.

foundation—with a southern exposure for solar energy in the winter. And all this was closer to our windplant than the cottage had been. In generating power from the wind, the closer the storage batteries are to the generator, the more efficient the system. Twelve-volt DC loses a lot of its power when carried over any distance, and batteries must be kept out of the cold to operate efficiently.

Although we knew roughly that our site faced south, we did not have a compass to get an exact orientation. Later, a few weeks into the project, when I was able to sight up the side of the east wall, I discovered that the house aligned perfectly with Polaris, the North Star. Luck, I admit, but good luck nonetheless.

The site had been part of a meadow when our homestead was part of a hilltop farm many years ago, but this particular corner of the meadow was now overgrown with small apple, wild cherry, and poplar trees. It took Jaki and me a couple of days just to clear enough of the growth to see the general terrain at the site. But we knew we would need a much more accurate understanding of the geography than we could discern visually. After all, since front-end loaders cost $22 an hour to hire, it is important to plan so that just the right amount of ground is moved, and not moved twice!

Our method of siting the corners worked extremely well for us, and I would recommend it to anyone faced with a hilltop or complex hillside situation. First we set up a surveyor's transit at the top of the knoll. We plumbed to a point on the ground within the legs of the tripod and marked the spot with a half brick. We set the zero-degree mark of the protractor to a known point (the corner of the cottage), knowing that the cave site was certain to fall within the semicircle described by the protractor. Then, with Jaki holding a calibrated stick and carrying one end of a 50-foot tape, and with me reading the level and marking the distances on a chart, we statistically mapped the area.

The procedure was to establish a slope for each ray of the semicircle divisible by 15°: 0°, 15°, 30°, 45°, and so on. I would set the level at 0°, for example, and Jaki would walk away from me with the stick until we were able to discern a drop in the land. The stick's calibrations began reading upward in inches from a point on the stick equal to the height of the level off the ground, say 4 feet. As soon as we could perceive an inch of drop, we took a measurement from the center of the half brick to the stick and recorded the distance. Then Jaki moved farther way with the stick on the same ray until we read a 6-inch drop. Again we measured the distance. Likewise, we recorded every 6-inch drop along the ray until we were outside the vicinity of the cave site. Then we repeated the procedure along the 15° ray, and so on to 180°. We then had a statistical abstract of the large general area of the building site (Table 1). This took us all afternoon.

I took the surveyor's level out of the open for the evening and went inside to transcribe the figures onto a large piece of graph paper. I let ¼ inch—the size of the squares on the paper—equal 1 foot. I drew a light pencil line at every 15 degrees of arc for the half circle and measured the scaled distances from the center point, marking with a dot each 6-inch drop in contour along the ray. To avoid confusion later, I lightly marked the elevation next to the corresponding dot, –1 inch, –6 inches, –12 inches, and so on.

We considered –1 inch level for our purposes. After all the figures from my chart were thus transcribed, it was an easy matter to connect all the dots of like elevations with gently curving lines. Lo and behold, an extremely accurate contour map of the site emerged as the dots were connected.

All this may sound like a lot of work for little gain, but it pays off by eliminating a tremendous amount of guessing and donkey work. From a piece of the same kind of graph paper, I cut out a scale model of the foundation plan using the outside dimensions of the planned block wall. I noted the location of the door on the south wall of the scaled drawing. Now all we had to do was slide the little square around the contour map until the most sensible location emerged.

We knew that the top of the block wall on the south, where there would be large thermopane windows, would be 42 inches below the top of the other three walls. We knew that the concrete slab at door level would be 78 inches below the tops of the three full-sized walls. As seen in Figure 5, the north and south walls are about even with the original ground level. The south wall required 6 inches additional excavation. The earth that comes out of the hole during excavation is used to build the terrain up to the tops of the east and west walls. By this plan, the new ground level melds nicely into the shallow-pitched, earth-covered roof.

Our job, then, was to plot the location of the excavation and formulate the most efficient depth of the excavation to take advantage of natural contours, keeping in mind that we would have to do something with every cubic foot of material that came out of the hole. Drawing an accurate contour map made this figure much easier to estimate. In retrospect, it was worth the effort: the landscaping around our cave did not turn out to be too big a job.

But there is another big time saver to come out of these little maps, now that the trouble has been taken to make them: the remarkable ease of placing the four flags to guide the excavation contractor. Once we determined the best location for the house on paper, all we had to do was to transpose the four corner points to the site itself. We used a simple angle-and-distance system to coordinate the points. Using a protractor and ruler, it was easy to determine, for example, that the northeast corner of the house should be 18 feet 3 inches from the half brick, at an angle of 29°. (This is why it is imperative to be able to set up the surveyor's level in exactly the same spot as it was when the original figures were taken.) We set up the level over the brick using the corner of the cottage as 0°, as we'd done before, and, at an angle of 29°, we measured out 18 feet 3 inches and drove a white birch peg into the ground (Fig. 6). We did the same with the other three corners. Then we checked the wall lengths and diagonals with our 50-foot tape, holding the tape level to account for the drop in the terrain. Every measurement was within 5 inches, which is very good. Ten minutes of juggling put all the pegs spot on the money. Walls were now the right length; diagonals checked.

I just happened to have a 4-by-4-foot sheet of plywood lying around, and I used this to establish the corners of the excavation itself (Fig. 7). Then we drove in another set of hefty birch pegs to mark these points. Of course, the four pegs marking the house corners would be eaten up by the front-end loader, but I wanted the operator to be able to visualize the project clearly before starting. I feel it is important for heavy equipment operators to see the site as the owner-builder does. With our contour map, it was easy to explain just how deep to go at each corner of the excavation.

A 4-foot work space all around the walls may seem like a lot, but it is better to have the room than to be crowded when it comes to building and waterproofing the walls.

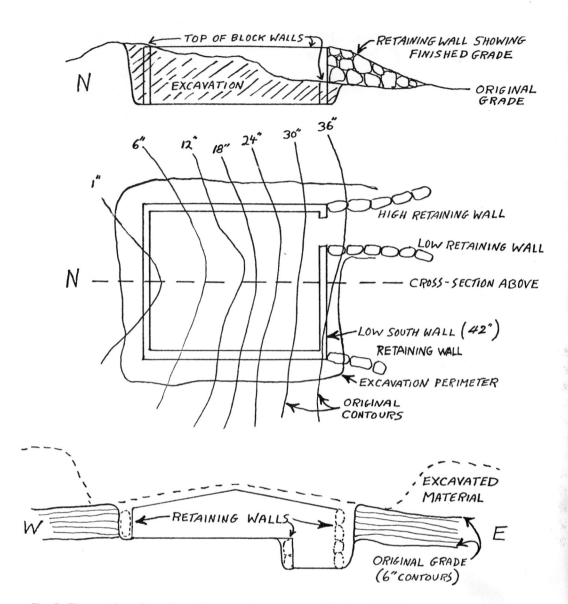

Fig. 5. The north and south walls were almost even with the original ground level; the terrain around the east and west walls was built up with excavated material.

Fig. 6. Transposing the four corner points from paper to the site, using a protractor, ruler, and surveyor's level.

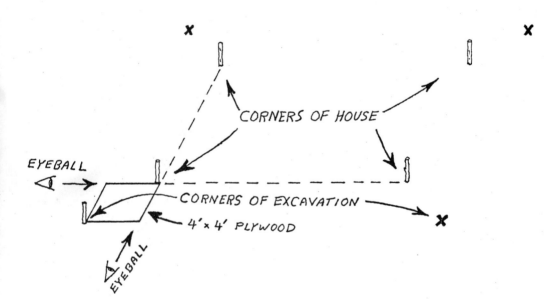

Fig. 7. Establishing the corners of the excavation.

A certain amount of erosion can be expected during the job, and even though we spaced our stakes 4 feet out from the corners in each direction, we did not end up with anything like 4 feet of space outside the walls. It was about 2½ feet, on the average. If the digger leaves the markers standing in the ground, the necessary slope will cut away into the 4 feet at the bottom (Fig. 8).

The reader may find this discussion of contours boring if the terrain near his building site is perfectly flat. For a flat terrain, I recommend the bermed type of subterranean house. Log-End Cave, in fact, is midway between the bermed and the chambered styles of building shown in Figure 1. With the bermed style, the builder need only calculate the amount of earth needed to shape up gracefully against the side walls. But keep in mind that unless the building site is entirely in sandy soil or good gravel, it is not wise to backfill with the same material that came out of the hole. Clay and other soils with bad percolation qualities should be kept away from walls that have to stay dry. Bring in sand or gravel, if necessary, to ensure good drainage. We had to backfill with twenty-five dumptruck loads of 5 cubic yards each because the ground in the vicinity of our homestead has such poor drainage. If it is necessary to bring in backfilling material, this should be figured in when calculating the depth of the excavation. If you are building a bermed house, it is easy to get rid of a little extra material. The thicker the berm, the better. In fact, if the earth piled up against the walls is thick enough—say, 5 or 6 feet— the heating advantages are almost the same as for a house just below the surface.

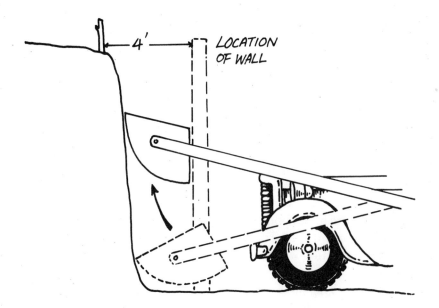

Fig. 8. A 4-foot work space all around the excavation is sufficient room for building and waterproofing the walls.

Let us calculate the required depth of excavation for the gabled berm-style house shown in Figure 9. We will assume poor soil percolation in the site vicinity, making it necessary to bring in backfilling material. If the site has good topsoil, the whole area should be scraped with a bulldozer and the topsoil piled up on a convenient side where it will be out of the way for the time being. This will save time and money later when the roof and final landscaping are done. We will not consider this hypothetical site blessed with a good depth of topsoil for the calculation, however. The reader may allow for it if pertinent at his site. The question, then, is: How much material of the kind taken from the excavation will be required on the berm? This problem could be solved with calculus, I'm sure, but not by me, and I can't explain something I don't understand myself. But we can get at the truth pretty quickly with some good old-fashioned "calculated guessing."

Let's say we were to excavate one yard deep over the whole area within a yard of the walls. This square, 12 yards on a side and 1 yard deep, will yield 144 cubic yards of material. But look at the berm. Because of poor percolation, the yard of space right next to the house will not be filled with the excavated earth. How much earth will the rest of the berm require? Well, the three sides of the berm directly adjacent to the sand backfill have the cross-sectional shape of a right triangle: 1 yard high (h) and 5 yards wide (b); that is, a cross-sectional area of $2\frac{1}{2}$ square yards ($A = \frac{1}{2}bh$, $\frac{1}{2} \times 5 \times 1 = 2\frac{1}{2}$). The total length of the berm is 34 yards ($11 + 12 + 11$) where it is directly adjacent to the sand backfill, so the cubic yardage is $2\frac{1}{2}$ times 34, or 85. Add to this the volume of the two delta-wing shapes where the berms meet at the corners, marked "D" on the diagram. The volume formula for those corners is $\frac{1}{4}hb^2$, or $\frac{1}{4} \times 1 \times 5 \times 5 = 6\frac{1}{4}$. In all, it will require $97\frac{1}{2}$ cubic yards of earth to build the berm ($85 + 6\frac{1}{4} + 6\frac{1}{4} = 97\frac{1}{2}$). But we've taken 144 cubic yards out of the hole! The 46.5 cubic yard difference is quite a bit to haul away or spread around the site.

Before we make a second guess, let's consider the situation if the excavated material had been of good enough drainage quality to use as backfill. The backfilling can be considered as a rectangular volume 1 yard wide, 2 yards high, and 32 yards long ($10 + 1 + 10 + 1 + 10 = 32$). The multiplication ($V = lhw$) yields 64 cubic yards. In this case, the total volume of the berm right up to the walls is $161\frac{1}{2}$ cubic yards ($97\frac{1}{2} + 64 = 161\frac{1}{2}$), a little more than the 144 cubic yards that came out of the hole. As loose earth always fills more space than it originally occupied, this isn't too bad. The berms could be made a little steeper, if necessary, or the excavation deepened slightly. Remember, too, that we need about 7 cubic yards as backfill along the front wall.

But we assume poor soil. Let's try excavating just $2\frac{1}{2}$ feet instead of 3 feet. This time, the volume of the excavation will be .833 of what it had been before ($2\frac{1}{2} \div 3 = .833$), or 120 cubic yards. The sand for drainage remains the same, 64 cubic yards. The berm is 6 inches higher now. The three in fifteen (3:15) pitch established by the roof adds 30 inches to the width of the berm at real ground level.* Expressed in yards, then, the berm is 1.167 yards high and 5.833 yards wide. The volume of the berm, as we saw in

---

* For ease of calculation in this example, I have drawn the gable one yard (3 feet) higher than the side walls, which are 5 yards (15 feet) away from the midpoint of the house. The roof pitch, therefore, is 3 feet of rise for 15 lateral feet, expressed 3:15. Roof pitch is usually measured in terms of rise per 12 lateral feet. Our example is equivalent to a $2\frac{1}{2}$:12 pitch. I recommend a pitch of 1:12 to 3:12 with an earth-covered roof, to minimize the dangers of earth sliding and soil erosion. Our own roof pitch is $1\frac{3}{4}$:12.

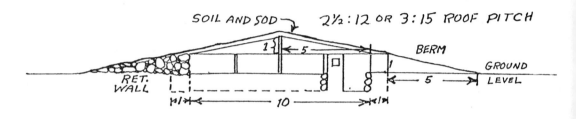

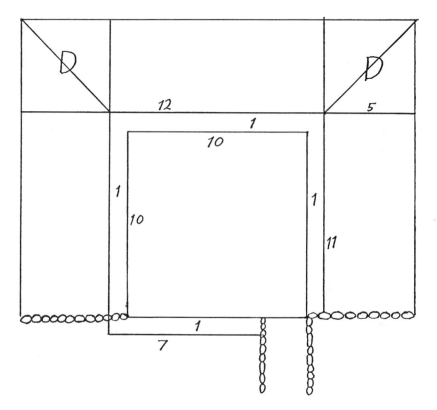

Fig. 9. Calculating the required depths of excavation for this hypothetical gabled berm-style house should take into account poor percolation and the necessity for extra backfill.

the first calculation, is $\frac{1}{2}$bhl (where 1 = length taken along the inside of the berm) plus 2($\frac{1}{4}$hb^2) for the delta-wing shapes where the berms meet. Substituting: (.5) (5.833) (1.167) (34) + (2) (.25) (1.167) (5.833) (5.833) = 115.72 + 19.85 = 135.57. We got 120 compacted cubic yards out of the hole. Not too bad. A shade deeper than 2$\frac{1}{2}$ feet should be perfect.

The example chosen is a realistic one. Such a bermed house would be just slightly smaller than the house we've built. The reader may think that a structure set only 30 inches into the ground can hardly be called subterranean housing. I won't argue that point, except to say that such a house is a much closer relative to subterranean living than it is to a conventional above-ground structure. The thick berm and earth-covered roof offer nearly the same advantages of heating and cooling as enjoyed by an underground dwelling. And the visual and environmental impact is about the same as that of our own house. From not too great a distance, the house would look like little more than a knoll on a flat landscape.

Consideration must be given to the disposal of waste water at the earliest stages of planning and siting. Waste-water disposal systems based on pumping waste water are all expensive beyond the kind of economic parameters described in this book. Moreover, pumping systems "are subject to various difficulties that beset equipment with regard to maintenance, the constant consumption of power over its lifetime, the problem of reliability, and nonavailability during power failure." [5] Therefore, I strongly recommend that most careful consideration be given to the relationship between the siting of the subterranean house and the location of the septic tank and drain field. On the side of a hill (the ideal site, especially if the hillside faces south), waste drainage by gravity will not be a problem. On a flat site, it may be necessary to dig deeper tracks and drain fields than normal, keep the elevation of the house higher (which may necessitate more landscape sculpturing to build the berms), or even raise the floor level of the bathroom by two or three steps in order to establish gradient for a gravity waste-disposal system.

If the water table at the building site is ever likely to be higher than the drains, rule out the site immediately. Even if the walls were as watertight as a swimming pool, the waste-disposal system would fail and probably back up into the house.

Malcolm Wells, writing in *CoEvolution Quarterly* (Fall 1976), says, "If the water table is high, build an artificial hill. When you see the way it bursts into life you'll know it's far less artificial than the stuff we've been throwing up all our lives in the name of architecture."

A final note on site location: Do not lose the advantage of the southern exposure by building behind evergreen trees. In northern latitudes, the sun is very low in the winter sky. Let its warming rays into the house. Trees that lose their leaves in the fall are not so much of a problem unless the site is in the middle of a thick forest.

3. Excavation

Together with backfilling and landscaping, excavation represents one of the biggest—and costliest—jobs connected with building underground. It is important that the owner-builder know what kind of equipment he needs as well as what it will cost him.

Estimates should be obtained from several heavy-equipment contractors. You can get an estimate for the whole job or you can pay by the hour for the machine. If the contractor knows his business and has a good reputation around town, the job estimate will be pretty darned close to the cost on a per-hour basis. Some contractors may try to tack on a hefty profit for the job. You can't blame them. But I always get a per-hour estimate as well. Make sure that you get all the facts and that the contractors are pricing for the same thing; otherwise no intelligent comparison can be made. For example, Contractor A gets $16 an hour for his backhoe. Contractor B gets $18 an hour. But Contractor A uses a backhoe with an 18-inch bucket, and B has a 24-inch bucket. For excavation, B is the better bet. Remember to ask if the contractor charges for hauling the equipment to the site. Many do. The firm I dealt with doesn't. This can make quite a difference, especially when there is a lot of follow-up work to do.

But price isn't the whole story. Ability is another. If two contractors gave me similar estimates, but one had a better reputation, I'd go with the good reputation, even if it cost me a few dollars more. I also prefer to pay by the hour rather than by the job. There are so many imponderables to subterranean housing. You may change your plans once or twice during construction, about where to put a soakaway, for example. If you're paying by the job, the contractor will penalize a change in plans.

What is the right equipment for the job? At the cottage, our cellar hole was dug entirely with a backhoe. It worked well, so naturally I assumed that I needed a backhoe for the cave. I did, and quite a bit, too, but the major excavation was done much more efficiently with a front-end loader. The loader was $22 an hour, the backhoe only $16, but the loader probably did the job in little more than half the time it would have taken the backhoe.

The excavation was much bigger than it had been at the cottage and, because it was cut into a hillside, it was easy for the front-end loader, with its 6-foot bucket, to maneuver. The entranceway on the south side of the house was the natural way for the loader to come in and out of the hole easily. It is true that the loader has to drive away with each bucketload and dump it and that a backhoe can stand in one spot and place the earth outside the excavation with its long boom. But the loader is moving darned near a cubic yard with each scoop. Actually, to finish some of the corners, we did use a backhoe instead; at that point the loader had to travel too far with each load to dump it where it would be useful for landscaping later.

The front-end loader would also do well for the excavation of the berm-style house used as an example in Chapter 2. It can make its own ramp down to the proper level, most conveniently at the place on the south wall where the front entrance will be. In a relatively shallow excavation without a lot of big rocks, a good man on a bulldozer can do a remarkable job, and bulldozers are cheaper to hire.

The backhoe, however, is the only machine for digging the septic lines, soakaways, and drainage ditches, but wait until construction is finished. It is dangerous and inconvenient to have ditches and holes all over the place.

In pricing for the excavation, bear in mind that you will need heavy equipment later for backfilling, landscaping, and perhaps a septic system. A contractor will be inclined to give a better price if justified by the volume of work. I stayed with the same contractor throughout the work at Log-End Cave. Consequently, I received excellent service as a regular customer.

4. Footing

The footing is the projected base of a wall. It supports the entire structure and distributes the weight of the walls over a large area. The footing should be reinforced so it is a single unit, a cast ring of the best concrete and iron reinforcing. Its dimensions follow a simple formula: the width of the footing should be twice the width of the block wall it is to support; the depth of the footing should be equal to the width of the wall. In our house, I decided on 12-inch concrete blocks, so the footings had to be two feet wide and a foot deep.

The site should be excavated to a flat surface about two feet beyond the outside of the planned footing (Fig. 11). Check the level with a surveyor's level and measuring stick. Because of the slope that will be left if the work is done with a front-end loader, the two-foot figure is consistent with the placement of the second set of flags or white pegs four feet outside the set of flags marking the house corners.

It is necessary to mark the location of the four outside corners of the footing so that a backhoe can draw the tracks within which the footing will be poured. There are two ways to plot these corners: batter boards and educated guesswork. Going by the book, Jaki and I built four batter boards way up on the surface so that we could slide strings back and forth to make sure the sides were the right length and square. These boards can be seen in Figure 11. Jonathon Cross, a local contractor and good friend, came along on a Saturday morning and found Jaki and me struggling with these grotesque, stick-figure batter boards. We'd been at it for hours. We figured that once we had the batter boards established, we could use them for the footings' inside and outside dimensions as well as for the block walls. "Don't need 'em," said Jon. "They'll just get in the way of the backhoe." We moved to the educated guesswork method then and there. It was so much easier that I'm not going to waste the reader's time with a detailed description of batter boards. Suffice it to say that batter boards are fine for work on the surface, especially on a sloped surface—say, to establish the four corners of a pillar-supported structure. When working around a huge, gaping hole in the ground, they're far more trouble than they're worth. Let this be the first of several mistakes we made that the reader can avoid.

Figs. 10. and 11. Excavating into a hillside is easier with a front-end loader and its 6-foot bucket. The site should be excavated to a flat surface about 2 feet beyond the outside of the planned footing.

When you've got a leveled area with dimensions 4 feet greater than the dimensions of your footing, drive a 2-by-4-inch stake at one corner (the northwest, for example), 2 feet in from each side of the flat area. Put a sixteen-penny nail in the top of the stake, leaving the head sticking out an inch for tying the mason's line. (You might as well go out and buy a ball of good mason's line; you'll be using it a lot on this project.) Now measure the length of the footing along the north wall to a point (35 feet 11 inches in our case), keeping about 2 feet in from the sloped edge of the excavation. Drive a stake into the ground and a nail into the top of it. It may be necessary to move this stake in a few minutes, as you will see. Now figure the hypotenuse (diagonal measurement) of your footing figures. We'll use our own figures as an example. Our footing dimensions are 30 feet $8\frac{1}{2}$ inches by 35 feet 11 inches or 30.71 by 35.92. With the Pythagorean theorem, we can calculate the diagonal measurement of the footing:

$$c^2 = a^2 + b^2$$
$$c = \sqrt{a^2 + b^2}$$
$$c = \sqrt{(30.71)^2 + (35.92)^2}$$
$$c = \sqrt{943.10 + 1290.25}$$
$$c = \sqrt{2233.35}$$
$$c = 47.258 \text{ Or } 47 \text{ feet } 3 \text{ inches}$$

These calculations can be done in a jiffy with a cheap calculator that has the square root function.

Now hook your tape to the nail on the northwest corner and, in the ground near the southwest corner, describe an arc of a radius equal to the shorter measurement of your footing dimensions (30 feet 8½ inches in our case). Next, hook the tape on the nail at the northeast corner and describe a second arc equal to your diagonal measurement (47 feet 3 inches in our example). The point where the two arcs intersect is the southwest corner. Drive in a stake and put a nail on top. Finally, find the southeast corner by intersecting the east side measurement with the south side measurement. Check your work by measuring the other diagonal. The diagonals must be the same in order for the rectangle to have four square corners.

You may find that the rectangle you have described does not use the cleared space to the best advantage. For example, you might find that you are crowding one of the excavation slopes, but have plenty of room on the adjacent side. It will not take long to rotate the whole rectangle slightly to alleviate this problem. You might even have to do a little pick and shovel work if one of the sides does not have enough room.

If you do not have a calculator and cannot figure the square root for your diagonal measurement, approximate your northwest, northeast, and southwest corners and make a mark at each place. Then intersect the east side and the south side and make a mark. Measure the diagonals. If they are, say, 12 inches off, add 6 inches to the lower figure and use that as your diagonal measurement. Adjust the northeast and southwest corners so you have the new diagonal measurement, and proceed as before.

Obviously, this calculated guessing will take a few trials to get all four sides and the two diagonals to check, but it sure beats making batter boards that you're only going to use for one job. In reality, we had our four corners to within a half inch in twenty minutes, and that's accurate enough for the footing.

To get ready for the backhoe, it is a good idea to place flags or white stakes on the various bankings for the operator to use as a guide. You can set these guides by eye-balling. Sight from one stake to the other and instruct a helper to plant a third marker on the banking in line with the two stakes. In all, then, you will have placed eight markers on the banks of the excavation for use as guides.

How deep to go with the trench for the footing is the next question to answer. This should be figured carefully. The important relationship is the one between the level of the top of the footing and the top of the floor. Look at Figure 12. The crosshatching represents undisturbed earth. For underfloor drainage purposes, it is good policy to bring in 4 inches of sand or gravel to lay the floor on. (This will be discussed in detail later.) I believe in a full 4-inch concrete floor. Ours, in fact, averages 5 inches thick and is not less than 4 inches at any point. Notice that the concrete floor is planned to catch the first course of blocks by 2 inches. This keeps the base of the wall from moving inwards due to the lateral pressure of the backfilling. In our case, the footing is planned to be 24 inches wide by 12 inches deep. Two inches of the floor, as well as 4 inches of compacted sand or gravel, will be below the top of the footing. This leaves 6 inches which must be excavated to accommodate the footing forms.

A 24-inch backhoe bucket is the best tool for the job. It may not seem too much to do by hand with a pick and shovel, but in tightly packed earth with large boulders (such as we had to deal with), excavation of footing tracks by hand would be a killer. In our case there was no decision to make, since the backhoe was still on the site. (It will save you time if you ask the operator to try to leave the four corner stakes undisturbed, even if you have to do a little pick-and-shovel work at each corner.)

If, in digging the footing tracks, a large boulder is encountered near the bottom of the track, wash it clean with water and a stiff brush and leave it as a part of the footing. Removing it would only be time-consuming and inconvenient; the hole left by such a boulder would have to be filled and compacted. It is best to pour footings on earth that has been left undisturbed. If earth is disturbed and replaced, it must be compacted before any concrete is poured over the area. This goes for the floor slab, too.

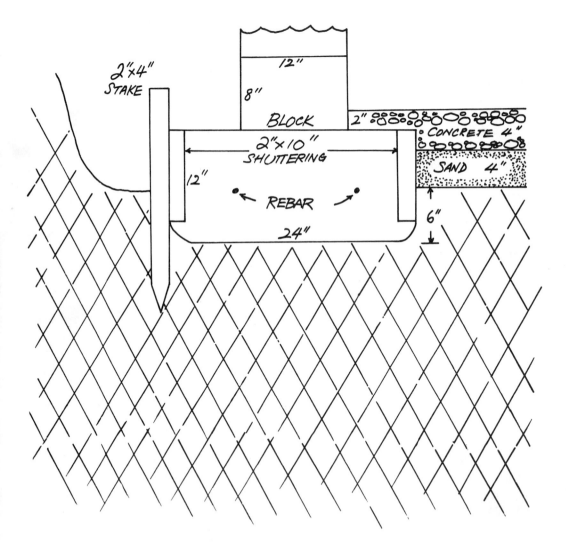

Fig. 12. The trench dug for the footing tracks of Log-End Cave was 6 inches deep.

Frostwall

In northern areas it is necessary to protect footings close to the surface from "frost heaving." Heaving can occur when wet ground below the footing freezes and expands, pushing upwards on the footing. The massive weight of the wall is of little help. The expansion forces of freezing care little how much weight they are called upon to lift. The only solution is to make sure that the ground below the footing will not freeze. This can be done by going deeper with the footing, protecting it with rigid Styrofoam insulation, or both.

The frostwall in the case of Log-End Cave is that portion of the south wall which is entirely above final grade, approximately 8 feet of wall near to and including the door. The footing below this portion of wall does not have the protection of 3 feet or more of earth above it, as is the case with the rest of the walls. So it is necessary to beef up the footing at this point to prevent frost heaving. We did this by going 2 feet deep with the footing over 12 feet of its distance in the southeast corner (Fig. 13).

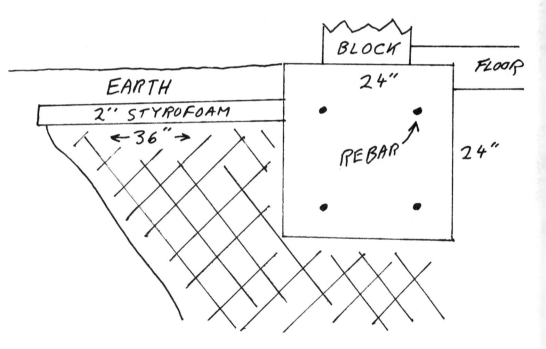

Fig. 13. Log-End Cave's frostwall is that part of the south wall which is entirely above final grade; part of its footing was dug 2 feet deep to prevent frost heaving. The Styrofoam just below the surface will insulate the ground in the cross-hatched area, protecting the frostwall.

Because of the sheltered nature of our entrance as well as its southerly exposure, I feel that a 2-foot frostwall footing is more than adequate, even though frosts of 4 feet may occur in unprotected areas in upstate New York. Remember that this is my opinion and though I believe it to be a sound opinion, others may differ. If in doubt, go deeper with your footing at the frostwall or protect the adjacent earth as shown in the drawing.

A 24-inch backhoe bucket will draw a track wider than 24 inches because the sides of the trench tend to crumble and it is difficult to keep the bucket moving in exactly the required path. This is fine. The extra room in the track will accommodate the footing forms.

We used 2-by-10-inch forms kindly lent to us by Jon Cross. This worked out well in shuttering for a 12-inch-deep footing. We let the concrete "leak" out of the bottom of the forms somewhat, giving a sturdy bell-shaped base to the footing. Obviously, you need to check to make sure that too much concrete cannot leak out in this fashion. It may be necessary to pack the outside of the trench with earth, stones, or even a 2-by-4. Do this well and double check it before the ready-mix truck arrives. Remember that concrete is heavy stuff, so these little dikes have to be sturdy indeed to stop the pour from breaking out of the intended location. Even so, have short boards and heavy stones ready to build a fast dike if necessary.

Two-by-12s are hard to come by, but even if they had been available, I think the 2-by-10s would have been preferable for ease of placement in an uneven trench. Two-by-12s would have required more space—and time on the backhoe—and would not have done a better job, in my opinion.

Preparing the forms for use is time consuming. First you must figure out exactly what lengths your forms need to be. No room for error here; you don't want to cut a long 2-by-10 2 inches too short. I drew a diagram of the whole shuttering system, showing clearly how the corners were to be constructed (Fig. 14). Note that the long forms on the north and south sides are 3 inches longer than the footing to allow the $1\frac{1}{2}$-inch-thick east and west side forms to butt against them. Conversely, the east and west forms on the inner ring are 3 inches shorter than the inner measurement of the footing because the thickness of the planks of the other two forms will make up the difference. All the other forms are consistent with the actual footing measurements. (The footing measurements given are based on the location of 12-inch blocks laid by the surface bonding method. This will be discussed later in detail.)

If the planking has cement on it from previous use, scrape it clean, especially those surfaces that will be in contact with the new concrete. Otherwise the shuttering will be hard to remove after the pour has set.

It will be necessary to cleat two planks together to make each length of shuttering (Fig. 15). Cut cleats 3 or 4 feet long and use 16-penny scaffolding nails to fasten the cleats to the planks. Scaffolding nails have two heads so that they can be easily removed when the shuttering is to be dismantled. I used ten nails on each cleat, five for each plank. Be sure to butt the planks tightly together and eyeball them dead straight with one another before nailing the cleat. Note that the required measurement for our north and south exterior shuttering is 36 feet 2 inches. To make good use of 18 footers, we cheated a little and left them 2 inches short of butting against each other. This is

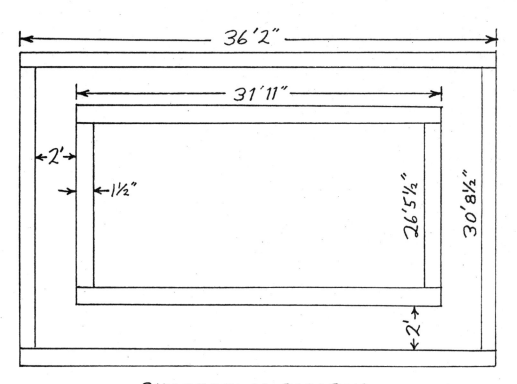

SHUTTERING DIAGRAM
OUTSIDE FOOTING DIMENSIONS: 35'11" × 30'8½"
INSIDE FOOTING DIMENSIONS: 31'11" × 26'8½"
DRAWING NOT TO SCALE

Fig. 14. Shuttering diagram showing the lengths of all footing forms.

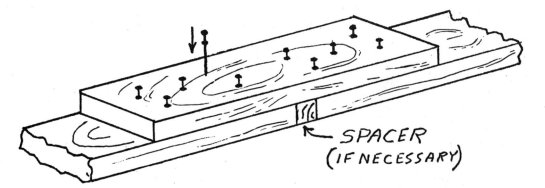

SPACER
(IF NECESSARY)

Fig. 15. Cleating two planks together.

okay as long as you nail a little 1½-by-2-inch spacer into the gap. The spacer will give almost the same stability as two planks butting directly against each other.

You will need a large number of 2-by-4 stakes; figure five or six for each length of shuttering and a few extra for the ones you smash to splinters with the sledgehammer. Fifty or sixty should do it for footings of a size similar to ours. Again, I was fortunate in borrowing stakes from Jon, but you can make your own from economy grade 2-by-4s. Fifteen 8-foot 2-by-4s will make sixty 2-foot stakes at a cost of $12 to $15. Put sharp tapered points on each stake.

Placing the Forms

This job would be extremely difficult without a surveyor's level. Beg, borrow, or hire one. The most important consideration in placing the forms is that they be level with one another. Set up the surveyor's level at some point outside the foundation which allows a clear view of a stick held at each of the four corners (Fig. 16). Using the existing corner stakes as guides, bring in the longer of the north side forms and put it roughly in place. Drive new stakes into the ground at the northwest and northeast corners, positioning them correctly so that they will be on the outside of the forms. Keep all the stakes out of the space where the pour is to be made. Drive the stakes so that the top of each stake is 6 inches above the average grade of the floor excavation. This allows for the 4 inches of compacted sand and 2 of the 4 inches of the floor. Remember that final floor level will be 2 inches higher than the top of the footing. To get the average grade of the excavation, take ten or twelve readings with the surveyor's level and measuring stick.

Fig. 16. The most important consideration in placing the footing forms is that they be level with one another.

Nail the long form to the two new stakes so that the top of the form is level with the top of the stake. Eyeball the form straight and drive a third stake into the ground about half way along. Only the corner stakes need to be level. Other stakes can be driven slightly lower than the level of the forms, so that they will be out of the way of screeding when the pour is made. "Screeding" is the term for flattening the top of the concrete so that it is even with the top of the shuttering.

Now level the form. This is a three-person operation: one to hold the measuring stick, one to read the surveyor's level, and one to pound the nails. Again, use scaffolding nails, coming in from the outside of the stake and into the forms. (A couple of tips: Drive your nails into the stake before you pound the stake into the ground, so that the point is just barely showing. Use a support stick or sledgehammer to help resist the pressure as you drive the nails all the way into the forms. Put at least one stake between the corners and the center, maybe two over an exceptionally long span. Check for level along the whole form.)

One down, seven to go. The rest are done in the same way as the first. Complete the outer shuttering before proceeding to the inner ring. Make sure the diagonals check! The inner ring is constructed to leave a space equal to the width of the footing (2 feet in our case). The inner ring must be level with the outer ring. Be sure to check the whole job, moving every 8 feet or so around the forms. Make slight adjustments by moving the stakes up or down. Use a lever to move stakes upwards. Finally, nail 2-by-4 buttresses every 10 feet or so to resist the tremendous outward pressure which will be exerted on the shuttering during the pour. Another means of resisting this pressure is by constructing movable cleats (Fig. 17).

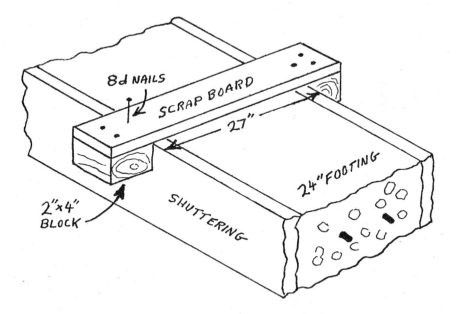

Fig. 17. Movable cleats will keep the shuttering intact during the pour.

Final Preparations before the Pour

Reinforce the footings with $\frac{1}{2}$-inch iron rebar, two or three bars throughout the footing, four or five in the frostwall portion. Place the bars around the perimeter, ready to fit. Overlap the rebar by a couple of feet and tie it together with wire. Bend right angles in some of the lengths for use at the corners. We found the perfect material for rebar in old silo hoops. The hoops were left over from the building of the cottage—we'd used an old silo as a major source of decking and flooring. If someone has torn down a silo in your area, try to work a deal for the hoops. They're great as strong, cheap rebar, but not much good for anything else if the threads are badly rusted. Usually, old hoops have to be cut apart with a hacksaw, and straightening them is a job, but the saving is well worth the effort.

Calculate the amount of concrete you need as accurately as possible. It sells by the cubic yard; the going rate in 1977 was about $32 for a cubic yard of the best concrete. Get the best, which is 3,000-pound mix, as opposed to 2,000 or 2,500-pound mix. Here are the calculations for our footings:

The volume of concrete equals the cross-sectional area of the footing times the perimeter. $V = 1' \times 2' \times 126' = 252$ cubic feet. Dividing by 27 gives cubic yards: $252 \div 27 = 9.33$ cubic yards. Add to this an extra cubic yard for the frostwall footing and the answer is 10.33 cubic yards. Actually, our computation was a little more involved than this since we were actually about 11 inches by 23 inches on the footing cross section. This reduced the figure to 9.3 cubic yards. The cement truck had a maximum capacity of 9.5 cubic yards, very close indeed to our estimate. So I said, "We'll chance it. Send up as full a load as you can carry." We ended up with about a wheelbarrow of concrete left over. Very close indeed. We were lucky. If possible, allow for a little greater margin of error than we did.

On the morning of the pour, apply a coat of engine oil to the inside edges of the forms. This makes removal and cleaning of the forms much easier. If you are borrowing the forms, you want to return them in as good condition as you found them.

Pouring the Footing

Round up plenty of help on the day of the pour, even Aunt Betsy. She'll be immensely valuable if she does nothing but keep the kids out of the concrete or the crew in hot dogs and beer. A crew of four or five is sufficient to draw the concrete around the ring, if they're fairly sturdy individuals (Fig. 18). Have four or five strong rakes on hand. A rake is the best tool for drawing concrete along between the forms. Put one or two folks in charge of laying the rebar. If that is their only job, it will not be forgotten in the panic.

Fig. 18. A crew of four or five, each with a good, strong rake, should be sufficient to draw the concrete around the footing ring.

Be Ready for That Truck!

They'll charge you overtime if the driver has to stand around and wait for last-minute preparations. Ask the driver for a stiff mix. The more water the easier the work, but the poorer the concrete. Sure, you could mix soup and it would practically flow around the shuttering by itself, but the footing would develop nasty shrinkage cracks. Mix stiff and work hard. Have ways cleared so that the cement truck can back up to different convenient locations to "shoot" the cement down the different tracks. A useful tool is a portable chute that can be set up on sawhorses or blocks to angle the concrete where you want it. Such a chute can be made out of half a sheet of plywood, cut lengthwise, with 2-by-4s for the sides. It is worth making this chute, and, if you wash it down after use, you will be able to use it again for the floor.

There's not much to say about drawing the concrete around the tracks except that it's darned hard work. The persons laying the rebar (Fig. 19) should try to get it in the middle of the footings, with 10 or 12 inches between the rods (see Figs. 12 and 13). After one side has been completed, one or two folks can start to "screed" the concrete before it sets up too much. Screed with a 2-by-4, drawing the concrete along with a constant backwards and forwards sliding motion (Fig. 20). You might use the screed on edge for the initial flattening, removing excess concrete as you go, and then give it another screeding with the side of the 2-by-4 to bring a little water up to the surface for a relatively smooth finish. Do not be concerned with too smooth a finish. You will be mortaring a course of blocks right on the footing, and the blocks will bind better to a slightly rough surface. If, after screeding, the surface becomes unduly smooth, you can scratch some roughness into it later in the day.

Have plenty of clean, washed stones the size of bread loaves to throw into the concrete in case you run a wee bit short. If you run very short, call for another truckload of concrete pronto.

The reader may have noticed that I do not even raise the possibility of mixing one's own concrete and building the footings—or, later, the floors—over several days. Yes, it would be cheaper. But it would not be as strong as a footing poured all at once. And it would be twenty times the work. The cost of the structure used as a model for this book is already so low that it does not pay to cut corners and possibly sacrifice strength.

Removing the Forms

Concrete dries at varying rates depending on its strength, initial moisture content, air temperature, and humidity. Remove the forms before the concrete has gone hard or you'll have a big job on your hands. If a corner of the concrete breaks off as you attempt to remove the form, wait a while longer. And don't neglect to clean those forms and stakes, especially if they belong to someone else.

As a point of interest, it took six of us two hours to draw and level the concrete. This is as it should be. You don't want to pay overtime on that cement truck. Jaki and I removed the forms the next day, breaking one which was held fast at a place where the concrete had leaked.

Fig. 19. The rebars should be laid in the middle of the footing, with 10 or 12 inches between the rods.

Fig. 20. The concrete can be screeded (made level with the top of the shuttering) with 2-by-4s.

5. The Floor

The reader will soon observe that the preparation for each job is almost always more work than the job itself. We poured our footing on July 11; we did not pour the floor until July 23; although we'd hoped to be ready by the 16th. It's just incredible how many things there are to do before the floor can be poured.

First, there are the 4 inches of sand to spread for good underfloor drainage. By July 14, the footing was aged enough to build a ramp of sand with a bulldozer over the frostwall portion. We brought in two loads of sand and dumped them just outside the "door." The bulldozer was then able to push and backblade the sand into the floor area. Using the footing as a guide for leveling, the dozer operator did an excellent job of spreading the sand evenly over the area, about 4 inches of sand throughout. He'd done all he could do in an hour, and we took our time finishing the job by hand, checking with the surveyor's level every once in a while. The next day I hired a powered compactor from the local rent-all store and tamped the sand dry. Jaki used a watering can to wet it and I tamped it again. This tamping is very important in giving the floor a solid base. A good rule is to compact any disturbed earth before pouring concrete over it.

Surface Bonding

Our 12-inch concrete-block walls are laid up by a masonry technique known as "surface bonding." This is a technique I would advise for other owner-builders. It is faster, easier, stronger, and no more expensive than conventional blockwork with mortar. The only alternative, in my opinion, is a poured and reinforced concrete wall, but that method would almost certainly require the services of a contractor skilled in poured walls and would be two or three times as expensive as the surface bonded walls described below, at least according to estimates that I have been able to get.

A solid concrete wall is stronger than a concrete-block wall of hollow-cored blocks, but a surface-bonded 12-inch concrete-block wall is already beyond the parameters of construction consistent with the safety of a structure of a size similar to the one described in this book. To explain surface bonding and its structural advantages, I can do no better than to quote from "Construction with Surface Bonding," by B. Carl Haynes, Jr., and J. W. Simons, Agriculture Information Bulletin No. 374, put out by the U.S. Department of Agriculture Research Service:

Surface bonding is both a material and a technique for erecting concrete-block walls without mortar joints. The bonding material is a cement-glass fiber mixture that is troweled on both sides of the stacked blocks to hold them together. No mortar is used between the blocks.

Normally, concrete blocks are laid in mortar. Contrary to popular belief, the mortar does not act as a "glue" to hold the blocks together. It serves mainly as a bed to aid in leveling the blocks. Mortar joints have little, if any, strength in tension and relatively poor adhesion. For structural purposes other than direct compression, the strength of a mortar joint is negligible.

Furthermore, mortar joints do not completely tighten the wall against the penetration of rain. The mortar joints actually serve as capillary wicks and draw moisture through the cracks between the mortar and the blocks. Being extremely porous, the blocks themselves also soak up water. Accordingly, a concrete block wall laid in mortar must be waterproofed by the application of a sealing compound.

In surface-bonded block walls, only the first course is bedded in mortar or bonding mix. This permits the accurate and rapid dry-stacking of subsequent courses.

Surface-bonded—or "skin-stressed"—concrete-block walls are stronger and tighter than conventionally laid walls. When the surface bonding mixture on the wall has cured, it will have relatively high tensile strength and good adhesion to the wall. Any flexure of a wall section is resisted to the limit of the bonding-tensile strength, and that strength is generally about six times that of conventionally mortared block walls. Also, the surface bonding mixture becomes a waterproof coating for the walls.

Another advantage of surface bonding is its economy: fewer hours are needed for wall construction, and less-skilled labor can be readily trained to apply surface bonding.

I strongly recommend that the reader obtain a copy of "Construction with Surface Bonding" if he intends to use this technique. It is available for $.45 from the Superintendent of Documents, U.S. Government Printing Office, Washington, D.C. 20402. Ask for Stock No. 0100-03340.

The First Course of Blocks

The footing dimensions of our house were designed to accommodate full courses of blocks without the need for cutting blocks. When using the surface-bonding technique the actual size of the blocks must be considered. Standard 16-inch blocks, for example, are really $15\frac{5}{8}$ inches long by $7\frac{5}{8}$ inches high. Table 2, which is from "Construction with Surface Bonding," gives true dry-stacked wall dimensions.

Table 2. Dimensions of Walls and Wall Openings Constructed with Surface-Bonded Concrete Blocks*

Number of blocks	Length of wall or width of door and window openings**	Number of courses	Height of wall or height of door and window openings***
1	1' 3$\frac{5}{8}$"	1	7$\frac{5}{8}$"
2	2' 7$\frac{1}{4}$"	2	1' 3$\frac{1}{4}$"
3	3' 10$\frac{7}{8}$"	3	1' 10$\frac{7}{8}$"
4	5' 2$\frac{1}{2}$"	4	2' 6$\frac{1}{2}$"
5	6' 6$\frac{1}{8}$"	5	3' 2$\frac{1}{8}$"
6	7' 9$\frac{3}{4}$"	6	3' 9$\frac{3}{4}$"
7	9' 1$\frac{3}{8}$"	7	4' 5$\frac{3}{8}$"
8	10' 5"	8	5' 1"
9	11' 8$\frac{5}{8}$"	9	5' 8$\frac{5}{8}$"
10	13' 0$\frac{1}{4}$"	10	6' 4$\frac{1}{4}$"
11	14' 3$\frac{7}{8}$"	11	6' 11$\frac{7}{8}$"
12	15' 7$\frac{1}{2}$"	12	7' 7$\frac{1}{2}$"
13	16' 11$\frac{1}{8}$"	13	8' 3$\frac{1}{8}$"
14	18' 2$\frac{3}{4}$"	14	8' 10$\frac{3}{4}$"
15	19' 6$\frac{3}{8}$"	15	9' 6$\frac{3}{8}$"

* Standard 16-inch blocks, 15$\frac{5}{8}$ inches long by 7$\frac{5}{8}$ inches high.
** Add one-fourth-inch for each approximate 10 feet of wall to allow for nonuniformity in size of blocks.
*** Make a trial stacking of blocks to determine the actual height of wall or opening before beginning construction.

"Construction with Surface Bonding" recommends adding $\frac{1}{4}$ inch for each approximate 10 feet of wall to allow for nonuniformity in the size of the blocks. Our block plans were based on size and space requirements of our general floor plan, as well as the availability and dimensions of certain materials, especially the roof framing timbers. The east-west length of Log-End Cave, then, is twenty-six whole blocks and one block laid widthwise, while the north-south dimension is predicated on twenty-two whole blocks and one laid widthwise (see Fig. 4). We can break twenty-six down to thirteen blocks twice and use the chart: 16 feet $11\frac{1}{8}$ inches $\times$ 2 = 33 feet $10\frac{1}{4}$ inches. Add 12 inches to this for the block laid widthwise and $\frac{3}{4}$ inch to allow for the nonuniformity of the blocks, and a total length of 34 feet 11 inches is obtained. Similarly, the east and west walls can be computed: 14 feet $3\frac{7}{8}$ inches (11 blocks) + 14 feet $3\frac{7}{8}$ inches (11 blocks) + 1 foot (the widthwise block) + $\frac{3}{4}$ inch (nonuniformity margin) = 29 feet $8\frac{1}{2}$ inches. The south wall is the same as the north wall, except that five blocks are left out where the door panel is located.

The first task in laying up the first course of blocks is to mark the corners. Use methods similar to those already described for establishing the corners of the footing. In theory, the outside edge of the block wall will be 6 inches in from each edge of the footing, but it is necessary to obtain $\frac{1}{4}$-inch accuracy, so fiddle with the tape and your marks until all the walls are the right length and the diagonals check. Now snap a chalkline between corners; the outside of the wall will be marked on the footing. Have a block plan (or use ours) and stick to it to get the first course of blocks in the right place. You will need to cut one 4-by-12-inch block to complete this and every course. A large block company may be able to supply you with these small blocks ready-made. The firm we dealt with could not.

This is as good a place as any to discuss blocks—the various sizes and materials, how they fit together, their cost, and so on. We chose 12-inch-wide concrete blocks: 12 inches for a stable wall, concrete for strength. Cinder blocks are lighter, cheaper, and easier to use, but for strength of construction they do not compare with concrete. The blocks we used have three full cores and two half cores in them, except the corner blocks which have the space filled where the half cores would normally be (Fig. 21). The regular

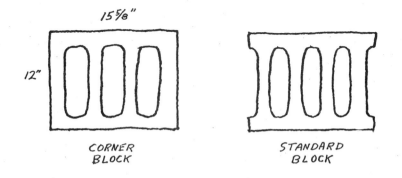

Fig. 21. Standard concrete blocks, weighing about 65 pounds each, have half cores at the ends; corner blocks are filled at the ends and weigh about 5 pounds more.

block weighs about 65 pounds, corner blocks about 5 pounds more. They are heavy. Solid 12-inch blocks would be killers to work with. Out local block company doesn't even offer solid 12-inch blocks. The other possibility is 8-inch blocks, maybe even 8-inch solid. But with 8-inch blocks, it is absolutely imperative to build a pilaster every 10 feet around the perimeter of the wall (Fig. 22).

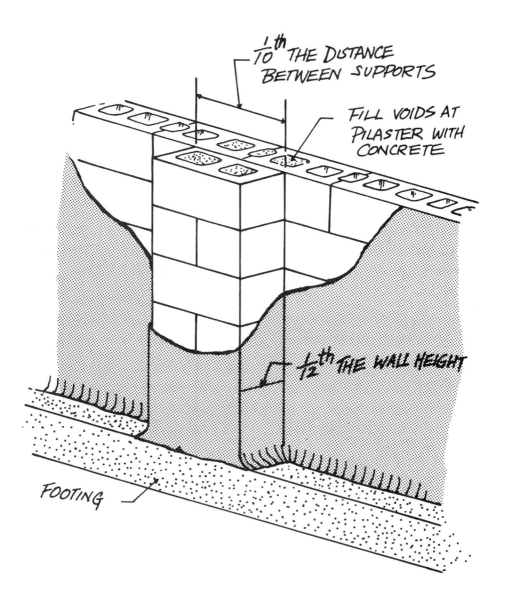

Fig. 22. A typical pilaster for strengthening a wall or supporting a beam. It is recommended if using 8-inch concrete blocks.

We used 8-inch blocks and pilasters in the basement at the cottage and grew to dislike pilasters. They get in the way of the floorplan, and we know from doing the corners at the cave that they would be a real pain to apply surface bonding to. Both in the laying up and in the surface bonding, they slow production enormously. Instead, we decided to go with the stable 12-inch blocks and filled the cores with rebar and concrete every place that would normally take a pilaster. In addition, we used the internal walls as buttresses against the lateral pressure of the walls. (I'll come back to these points later.)

Ten-inch blocks are also offered by most block companies, but I can't recommend them. Their awkward dimensions mean constant cutting of blocks, and you would still need those pesky pilasters. All blocks, by the way, are $15\frac{5}{8}$ inches long and $7\frac{5}{8}$ inches high.

Another consideration when weighing the 8-inch block and pilaster method against the 12-inch block method is that most blocks are made with mortar joints in mind. Remove the mortar joints, and corners and pilasters will be uneven with 8-inch blocks.

Twelve-inch blocks should be used with a block plan similar to our own to avoid cutting blocks. Cutting blocks is a downer, and a little adjusting of the dimensions of your house will save you all that effort. This is why we end each course with a block running widthwise, instead of having a whole number of blocks on a side. Consider Figure 23. A simple point perhaps, but a little adjustment on paper when you draw your plans can save days of work and lots of moaning later on.

As for cost, ask for a volume discount. You should get at least 10 percent off for five hundred or more blocks. Remember to ask about haulage charges, too. These charges usually vary with distance, but might also be a function of volume. If you are fortunate enough to have two or more suppliers, get comparative estimates. But also examine the blocks carefully. With surface bonding, it is worth a few cents extra for blocks of uniform dimensions. Banana blocks at $.10 off per block are no bargain. In 1977, we paid $.51 per block, a good price for good blocks.

According to "Construction with Surface Bonding": "Lay the first course of blocks in a rich mortar mixed by volume as follows: masonry cement, 1 part; portland cement, $\frac{1}{2}$ part; and sand, $2\frac{1}{2}$ parts. Use a stiff mixture if a thick mortar bed is needed to level the first course." To avoid dealing with half measures, we can double up on all the materials: 2 masonry, 1 portland, 5 sand.

Start with the highest corner, but know how much higher it is than other parts of the wall, so you can estimate a practical thickness for the mortar bed. Remember, the purpose of the first course is to establish a dead-level base for the dry-stacked courses, so the thickness of the mortar bed will vary around the building. My advice is to set the four corner blocks so that they are all at precisely the same level. Then tightly stretch a mason's line from the top of one corner to the top of the next. Stretch it as tightly as you can and tie it to something that won't move, like a sixty-five pound block. Eyeball the line. There will probably be a sixteenth of an inch dip. Keep that in mind as you proceed along the course, so that halfway along, the block is, say, a string's width above the line. Check each block for level as you lay it. The string will tell you the level along the length of the wall, but use a 2-foot level to check along its width. This is important. Laziness here will cost you time later. I know from experience. On our north wall, I allowed quite a lisp to develop in the first course, and it took considerable shimming later on to correct the problem.

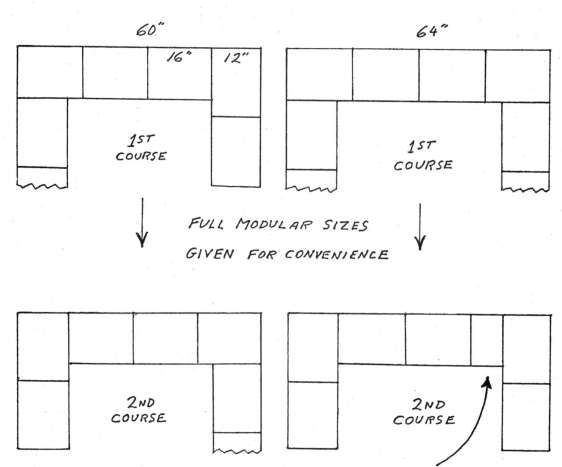

60"

16"   12"

1ST
COURSE

64"

1ST
COURSE

FULL MODULAR SIZES
GIVEN FOR CONVENIENCE

2ND
COURSE

2ND
COURSE

CUT BLOCK REQUIRED
EACH COURSE, EACH WALL

NO CUT BLOCKS REQUIRED HERE

Fig. 23. It's a good idea to adjust your plans a little so that corners can be made without having to cut blocks.

Fig. 24. Keep the blocks tight against each other.

Keep the blocks tight against each other (Fig. 24). Make sure no mortar has crept in between the blocks from below. A light tap with a hammer will usually assure that two adjacent blocks are well butted against each other. When you lay the last block, you may find that it does not fill the last gap exactly. If there is $\frac{1}{8}$-inch gap or less, that's fine. Over an eighth and you should bring the corner block in to fill the gap. You might even have to move the corner block out a shade to make room for the last block. I found it useful to check the length of the first six blocks I laid, to get an idea of how my work was comparing to the values given on the chart. Any adjustments in the quality of the work could then be made. Remember that we allowed an extra $\frac{1}{4}$ inch for each 10 feet of wall. I found this to be a generous allowance, but another company's blocks may be slightly different.

I'd never laid blocks before, though I'd labored for masons, but it only took about ten hours to get a decent job—except for the aforementioned north wall where I didn't use the level often enough.

In our house, it was necessary to set a 4-by-12-inch block next to the last full-sized block on the south wall to establish the base for a short pillar, which strengthens the end of that wall (see block plan, Fig. 4).

This first course of blocks establishes the edge of the concrete floor and provides a flat surface for use in screeding the floor. No further blocks will be laid until the floor is finished. In fact, the main part of the blocks should not even be delivered until the floor is hard, so that the pallets of blocks can be set right down onto the floor for easy access during the wall-building. We hauled the first ninety blocks for the first course with our little pick-up truck (three loads, and they are heavy).

There is much to do to get ready for pouring the floor: pillar footings, underfloor drains, understove air vents, rough plumbing, wire mesh to be cut and fitted for reinforcement.

Pillar Footings

The three main support beams at Log-End Cave are well supported by posts cut from 8-by-8-inch and 9-by-9-inch old barn beams. Each of these posts must have a pillar footing beneath it to support the heavy roof load without cracking the floor. We decided to make the footing under each post 2 feet square by 1 foot deep. We measured for the location of each footing, scaling off the rafter plan, and marked the squares in the sand with a stick. Because the level of sand was already designed to carry a 4-inch floor, we only had to excavate an additional 8 inches to give us the required 12-inch depth. Pick and shovel work. We hauled the material away with a wheelbarrow, though we used some of the sand to build up one or two low spots on the sand pad. It was stiflingly hot the week we prepared for the floor pouring, and the dogs found the pillar-footing holes to be ideal spots for lapping up some of the cooler earth temperature below the sand. These pillar footings are poured at the same time as the floor (Fig. 25).

Fig. 25. Pillar-footing holes should be poured at the same time as the floor.

Underfloor Drains

One theory formulated to protect a floor from rising dampness is to lay a plastic vapor barrier down before pouring the floor. We had done this at Log-End Cottage. But it's impossible to lay a perfect barrier. It will be ripped considerably by wire reinforcing and a dozen feet running all over it during the pour. A better idea, I think, is to put a good drainage material like sand below the floor so that water won't be inclined to rise at all. To expedite removal of any accumulation of underfloor water that might occur, we snaked perforated 4-inch drain tubing (also called "tile") around the floor area and took it under the south wall footing for later connection to a soakaway. This underfloor drain can be seen in Figure 26.

Fig. 26. Perforated 4-inch tubing was laid to remove any water that might accumulate under the floor.

Understove Vents

Woodstoves need air or they will choke from lack of oxygen. We laid 4-inch nonperforated flexible tubing under our floor and took it up through the floor under the planned location of the two main stoves. These two pipes connect with each other and travel under the footing to the lower of the two retaining walls near the front entrance. The outlet is covered with an insect and rodent-proof vent cover. We believe that this method of supplying air directly to the stoves from the outside prevents drafts and keeps the stoves from burning up all the oxygen in the house. The British have a similar method of bringing outside air to their fireplaces. I figured, if it works for a British fireplace, why not for an American stove? We believe the experiment has been a success.

Rough Plumbing

Now is the time, the only time, to install the meat of your drainage system. Ours is a simple system. The main sewer line to the septic tank is a 4-inch pipe. In the house, it is normal to tie all the fixtures to a 3-inch line, which expands to a 4-inch line outside the house. The toilet flushes directly into the 3-inch pipe. Other fixtures, such as bathtubs and wash basins, have a $1\frac{1}{2}$-inch outlet pipe, which joins the 3-inch pipe. A vent stack, necessary for proper operation of the drainage system, can be a $1\frac{1}{2}$-inch pipe. Its purpose is that of a pressure release to allow easy flow to the septic tank. Finally, we installed a clean-out just in front of the toilet and hid it beneath the floor, enabling us to run ramrods straight through to the septic tank from the bathroom if ever it should be necessary. The whole system can be seen clearly in Figure 27.

Make sure you have a slight slope towards the septic tank on all the lines, and try to place the toilet receptacle fitting exactly at the planned floor level. Leave other pipes sticking up a good foot above floor level, wire them to iron pipes so that they can't be moved accidentally, and cover the exposed openings to keep concrete from getting into your drain system. Most of our plumbing is in the bathroom and consists of a bathtub drain, a wash-basin drain, a vent stack (all $1\frac{1}{2}$-inch plastic pipe), a toilet receptacle, and a clean-out (both 3-inch plastic pipe). There are all sorts of different connectors, elbows, expanders, and clean-outs on the market. Find a hardware store or plumbing supply shop with a good stock of parts and a clerk who knows his stuff. You may have to pay a slightly higher price for parts, but if you get good sound advice along with them, it's worth it.

The only other plumbing at Log-End Cave is the kitchen sink outflow pipe. For convenience, we took this pipe under the footing and into the same soakaway that deals with our underfloor drains. This is an acceptable practice in most rural areas, but you should check with the county, town, or state health department, whichever has jurisdiction over your area on these matters. This simple gray water disposal system should not be used with garbage disposals.

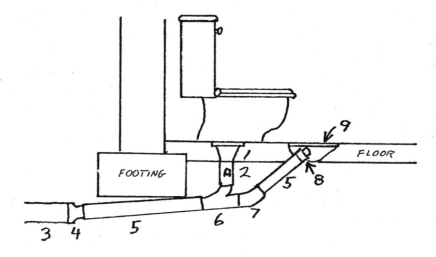

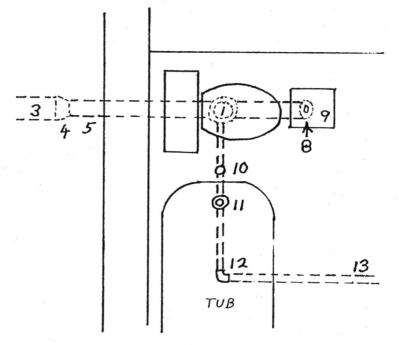

1. toilet receptacle unit
2. $1\frac{1}{2}''$ to 3" right angle join
   (from tub, vent, and basin)
3. 4" sewer pipe
4. 3" to 4" expander
5. 3" PVC pipe
6. 3" right angle join
7. 3" 45° angle
8. cleanout (3" threaded cap)
9. wooden cover
10. vent stack to outside
11. tub drain
12. $1\frac{1}{2}''$ PVC right angle
13. $1\frac{1}{2}''$ PVC pipe to basin

Fig. 27. The septic system.

Wire Mesh Reinforcing

The floor will be greatly strengthened by the use of welded wire mesh reinforcing, available at building supply yards. The mesh is commonly 6-by-6-inches in No. 10 or heavier gauge wire. We used "goat fencing" of similar dimensions for our reinforcing; hundreds of feet of it ran through the woods on our property. Apparently, someone many years ago had made some sort of corral in the woods. All we had to do was cut it and drag it out. The reinforcing can be seen clearly in Figure 28.

Fig. 28. Wire mesh (or "goat fencing") can be used to strengthen the floor.

The Fence

All that remains to prepare for the pour is to build a "fence" down the middle of the floor area for supporting the screeding planks. Drive 2-by-4 stakes firmly into the ground and nail a row of 2-by-4s to these stakes so that the top edge of the 2-by-4s is at the same level as the top surface of the first course of blocks. Cut off any stakes sticking up above the top edge of the fence. Then prepare a couple of screeding planks by notching out the appropriate amount of a 2-by-10 (6 inches in our case). We used the same 2-by-10s that were used as footing forms. The completed fence and the screeding planks can be seen in the photographs of the floor pouring (Figs. 29 and 30). The fence should be checked for elevation with the surveyor's level.

Finally, calculate the floor. With a 4-inch floor, the cubic yardage will be one-ninth of the square yardage (4 inches ÷ 36 inches = $\frac{1}{9}$). In our case, I knew that we would average close to a 5-inch thickness and figured accordingly: 5 inches ÷ 36 inches × 101 square yards = 14 cubic yards. We added 2 cubic yards for the pillar footings, giving 16 cubic yards. We knew it would require two loads to bring the concrete so we ordered 9 cubic yards on the first load, in case our estimates were a little off. Nine yards managed a little more than half the floor. Eight yards on the second load finished the job almost perfectly. Again, there was about a wheelbarrow load left over. So we actually used about $16\frac{3}{4}$ cubic yards, instead of the calculated 16, about 5 percent more. I think it is reasonable to figure 5 to 7 percent extra concrete to allow for error.

Fig. 29. As your pour the concrete floor, keep the wire mesh level.

Fig. 30. A power screed can be used on a shorter span of concrete, but in this case two men screeding with a single plank did just as well.

Pouring the Floor

I hope the reader does not experience the last minute rushing around that we had to go through. We were up at 5:00 A.M. finishing off the plumbing, and we finished leveling the fence about noon, while the mixer was mixing.

Get the same half-dozen helpers you rounded up for the footing pour, if they're still friends, and tell them you've really got a job for them this time. Experienced concrete pushers are hard to find. Again, for a strong floor, ask the driver for a stiff mix. Warn your helpers to be careful about raising the wire mesh (Fig. 29). It should be drawn up slightly with the tines of the rake as you pour. Best keep it within an inch or two of the bottom, though. This provides good strength and lessens the danger of getting the power trowel blades caught up in the mesh later on. The wire mesh should stay at the same level, even where it passes through the area of the pillar footings.

Do one side of the fence at a time. As soon as you've got a fair area poured—say, 20 percent of one side—start screeding. We used a power screed, which is nothing more than a big vibrator to which two wooden planks are attached. It was only moderately successful. I think our planks were too long. To draw the screed along required another man at each end of the planks helping to slide them along the fence and the first course of blocks. The power screed might have been better if used on a shorter span or with smoother supporting runs—not blocks—but we found out later that two men with a single plank could do as well (Fig. 30), saving a man's labor (and the cost of hiring the power screed, if we had known).

Water can be drawn to the surface and the floor smoothened further with a "bull-float," also shown in Figure 30. A bull-float is a large, perfectly flat masonry float, usually made of aluminum or magnesium. Extension handles can be added to avoid walking on the wet concrete. I borrowed a bull-float from Jon, but they can be hired at low cost from a tool rental store.

After the first pour, there was a good hour's wait for the truck to make its return. We used the time to catch up on screeding and to have lunch. After a while, though, the welcome rest became worrying. The first pour was setting rapidly and there was still no sign of the second load. We covered the first pour with Styrofoam sheets in an attempt to retard the setting, watered the edge where the two pours would join, and inserted 16-inch long iron reinforcing rods 8 inches into the edge of the first pour to help knit the two pours together. At last the truck arrived and we were back at work, refreshed by the break. Still, it was a little scary; if the second load had been delayed another half-hour, it might have affected the quality of the join. The moral: specify two trucks. Usually, the trucks have a radio in them, but if not, use the nearest phone to shout for the second load as soon as you can make a good estimate of what is needed to finish the job.

It was a long day. Drawing some 17 yards of concrete with rakes, screeding it, battling with wire mesh, working around plumbing and understove vents . . . it's tough. And you, the owner-builder, are responsible for the success or the failure of the project. It is you who must coordinate the various jobs, make the decisions, run around like the proverbial chicken. I was extremely fortunate to have my contractor friend Jon Cross on hand to keep us right. Jon's experience and take-charge attitude probably made the difference between a so-so floor and the excellent floor we have. If you can enlist the help of someone who has experience with concrete slabs, jump at the opportunity. Another time, someone might ask you for the benefit of your experience.

By five o'clock, we were ahead of the work and everyone went home except Jon Cross. We had supper and took it easy for an hour, Jon checking the set every few minutes to see if we could commence power troweling. He impressed upon me the importance of experience in the handling of the power trowel, a machine which can run away with the operator if one of its four rotating blades digs into the concrete. I was thankful that Jon was willing to run the machine for us. I labored for him by bringing water or cement as needed, and working the edges with a hand trowel. One trick in obtaining a smooth floor is to scatter a dusting of dry Portland cement on the floor. The power trowel then does an excellent job of producing a flat, smooth floor, and I highly recommend its use, which I cannot say for the power screed. We finished after 10:00 P.M., by car headlights.

The next day was a Sunday, and, though not normally our custom, we made sure it was a day of rest.

6. External Walls

The external walls can be made of stone, reinforced concrete, mortared blocks, or surface-bonded blocks. Building with stone means intensive labor. Three families living in our community have built cellars of stone. All three cellars were slightly smaller than our subterranean house. One of the men had masonry experience. Building with shuttering on the outside, free-form inside, he completed the cellar in three weeks, but he worked like a man possessed. The other two families spent entire summers on their cellar walls, and I doubt if either of them would meet waterproofing requirements for a permanent underground living space. There is still some argument as to the economy of building with stone instead of blocks. Pat and Joe, using the slip-form method, used something like two hundred bags of cement. They hauled the sand in from the community sandpit for free. Stone could be considered cheaper only if the owner-builder places a zero dollar value on his labor. And stone is much tougher to waterproof effectively. The outside surface of the wall must be parged (coated) with a smooth layer of cement plaster, and then waterproofing must be applied as if dealing with a mortared block wall.

Poured concrete walls are the common method employed by the well-known advocates of subterranean housing, such as Malcolm Wells, John Barnard, and "Caveman" Andy Davis. My feeling is that the shuttering involved is hardly worth the gain in structural strength. It may be cowardly on my part, but I think that a poured wall is better left to experienced contractors, which adds a lot to the cost of the home. It may well double or triple the cost of the external walls. If our walls had been of solid concrete, they would have cost us $1,437 in concrete alone. Add to this the cost of the shuttering, reinforcing rod, waterproofing, and labor, and we might well be approaching the $3,000 figure. By comparison, our 12-inch block walls cost us $901 in materials, inclusive of surface-bonding mix. I also paid a friend about $150 for labor on the job. Conclusion: poured concrete is stronger and more expensive. In my opinion, it is not that much stronger to justify the extra cost. Also, the other advocates of subterranean housing report that a poured wall takes a year or two to cure, which means a fairly damp atmosphere inside for quite some time after building.

This leaves blocks. My experience is that a mortared wall and a surface-bonded wall cost about the same if all factors, especially waterproofing considerations, are weighed. The mortared block method requires a lot more skill than surface bonding, and the surface bonding will yield a much stronger wall, as discussed earlier. My advice is to use 12-inch concrete blocks in conjunction with surface bonding. This method is stronger, more waterproof, and easier than a mortared wall, and a whole lot cheaper than a poured wall. At the same time, they maintain a good safety margin of strength and avoid a lot of the curing problem. Also, surface bonding is best for the inexperienced owner-builder. Thus it is the method described in this book. If the reader is not convinced by my argument, there is plenty of literature in good libraries about slip-form stone wall construction, block masonry, and poured-wall building.

My nephew Steve Roy and his friend Bruce Mayer arrived at Log End during the surface bonding operation. They were to become a welcome and indispensible addition to the work-team during the next three weeks.

Building Surface-Bonded Walls

The first course is already in place and level. I'll let "Construction with Surface Bonding" take up the story. My paraphrasing would not improve on that excellent text, but I have substituted photographs from our own work at the cave:

Lay succeeding courses of blocks without mortar. The tops of most blocks are somewhat rough. Slide the blocks back and forth a couple of times over other blocks to knock off excess material and burrs before stacking them. [Note: As seen in Figure 31, we used a piece of broken block for this purpose.]

Fig. 31. A piece of broken block can be slid across other blocks to knock off excess material and burrs before stacking.

Stack the blocks three courses high at corners and plumb. Stretch mason's line between corners. Then fill in between corners with blocks. Repeat the procedure with another three courses of blocks. [Figs. 32 and 33.]

Blocks that are not dimensionally true may be plumbed and leveled by inserting flat sheet-metal shims or brick ties between them. *Do not use wood shims or spacers.* If the height of the blocks varies by more than one-eighth inch, bed the short blocks in mortar or bonding mix to correct the excessive height difference. [Fortunately, our blocks were good and we never experienced such a variation.]

COMMERCIAL SURFACE BONDING PREMIXES. Dry, premixed surface bonding is now being packaged by a number of firms and is available on the retail market. Building material dealers, concrete products plants, and paint stores will be stocking these products as they become more widely used. The bag sizes range from 25 to 80 pounds. Some premixes require only the addition of water. Others have a small plastic envelope filled with calcium chloride inside the bag. The calcium chloride should be mixed with water before making the wet mix. Some premixes contain sand and should be applied one-eighth inch thick. Those without sand need be applied only one-sixteenth inch thick. Follow the manufacturer's directions on the package.

The price of commercial premixes may run as much as three times the cost of ingredients for home mixing. However, the commercial products are accurately proportioned and eliminate most of the labor of mixing. They also eliminate the need to locate ingredients, some which are sold only in large quantities.

Inspect the premix before using it to be sure that the fibers are well distributed; if they are not, remix. Also, inspect for frayed glass strands. If there is a large amount of frayed strands in several of the packages, do not accept the commercial premix. It will be difficult to mix and apply. Try to get packages from a different batch, as manufacturers are experimenting to get proper control in mixing their product.

Remix as little as possible. Too much mixing frays the strands or separates the strands into individual filaments. This makes proper application of the bonding mix difficult.

INGREDIENTS FOR HOME-MIXED SURFACE BONDING. The ingredients for home-mixed surface bonding are as follows:

● Portland cement (normally packaged in 94-pound sacks). White cement is more expensive than regular gray cement but is less alkaline, has a more finished appearance, and needs less mineral coloring for pastel shades if you desire to color the mix. It is preferred for all uses, but regular type I gray cement is sometimes used.

● Hydrated lime (normally packaged in 50-pound sacks). Hydrated lime makes the mixture more workable and easier to apply. Lime with lowest alkaline content is made from pure dolomitic limestone.

● Calcium chloride (normally packaged in 100-pound sacks), in flake or crystal form. Calcium chloride makes the mixture set up quicker and results in a harder surface. It is available from agricultural chemical dealers and from distributors handling it for ice and snow removal.

● Calcium stearate (normally packaged in 50-pound sacks). Calcium stearate makes the mix waterproof. Use a wettable technical grade, generally available from chemical distributors.

56

Figs. 32 and 33. After the first course of blocks is in place and level, build your surface-bonded wall by stacking the blocks three courses high at each corner and then filling in between the corners.

● Glass fiber filament chopped into one half-inch lengths (normally packed in 40- or 50-pound boxes). Type E fiber, coated with silane or chrome organic binder, is available from plastic and chemical supply distributors. An alkali-resistant fiber, type K, may be available from building material dealers and plastic products dealers. The glass fiber acts as reinforcement in the mixture to give it strength and prevent cracking.

HOME MIXING THE MATERIALS. The bonding mix sets rapidly after the water and calcium chloride have been added to the dry ingredients, especially in hot weather. If one person is plastering, prepare only 25 pounds of bonding mix at one time.

The weights of the ingredients needed to make a 25-pound batch (dry weight) of the bonding mix are as follows:

Ingredient	Parts	Pounds
Cement	78	$19\frac{1}{2}$
Lime	15	3-3/4
Calcium Stearate	1	$\frac{1}{4}$
Glass Fiber	4	1
Calcium Chloride	2	$\frac{1}{2}$
Total	100	25

Mix—in dry form—the cement, lime, and calcium stearate thoroughly. Add the glass fiber and remix only long enough to distribute the fibers well. Too much stirring tends to break up the strands into individual filaments. When this happens, the bonding mix is hard to apply . . . .

Mix the calcium chloride with 1 gallon of water. Add this solution slowly to the dry ingredients and mix thoroughly. Add about one-half gallon more of water. You may need to adjust this amount of water slightly to produce the right consistency for good troweling. The mix should have a creamy consistency—as thin as possible but not too thin to prevent handling with a trowel. Most people tend to make it too stiff. It will then be hard to apply and may not bond properly.

Mixing can be done by hand in a wheelbarrow or small mortar box. A garden cultivator rake or weeding hoe (three- or four-tine) works best. Check the mix with your hands for lumps. Wear rubber gloves to avoid possible burning of the skin . . .

If the mix becomes too stiff before it can be completely used, add a small amount of water. Do not add water more than 30 minutes after the initial mixing because it weakens the bond. Discard such remixed batches whenever the material again becomes too stiff to apply on a wet wall.

Batches of the dry ingredients can be mixed well in advance so that there will be no delay in preparing the mix when it is time to begin the bonding operation. If the dry mix is to be stored several weeks, place each batch in a plastic or multiwall paper bag and close the top tightly. Weigh out the calcium chloride for each batch and seal it in a separate plastic bag; do not mix it with the dry ingredients.

APPLYING THE BONDING MIX. *Surface-bond both sides of the wall.* It will not be strong enough if the bonding mix is applied on only one side.

The blocks must be free of dirt, loose sand, cement, and paint. If necessary, clean the blocks with a wire brush when they are dry.

Spray the wall with water until it is wet but not dripping [Fig. 34].

Fig. 34. Before applying the surface bonding, spray the wall with water until it is wet but not dripping.

Work the mix from a hawk onto the wall with a plasterer's trowel [Fig. 35]. Hold the hawk against the wall to avoid excessive spilling of the mix.

A very thin coat—about one-sixteenth inch thick—of the bonding mix is all that is necessary.

Work from the top of the wall downward. Thus, if the uncoated portion of the wall needs rewetting, the water will not run over freshly applied bonding.

Most workers can cover a section about 5 feet wide standing in one position. Start applying the bonding 2 or 3 feet from the top of the wall and trowel the mix upward to the plate. Move down another 2 or 3 feet and repeat the process, blending the freshly covered section into the bottom of the section above.

There are four essential steps in successfully applying and finishing the bonding:

1. Apply the mix with firm trowel pressure, pushing the load upward and outward until a fairly uniform coverage is attained.

2. Follow with longer, lighter strokes, holding the face of the trowel at a very slight angle to the surface (about 5°) to even up the plastered area and to spread excess bonding mix to fringe areas.

3. Move to the area below and apply mix as in steps 1 and 2. Continue bonding for 15 to 20 minutes, or until you have covered 25 to 30 square feet of surface.

Fig. 35. Work the mix onto the wall with a plasterer's trowel.

4. Dip the trowel in water to clean it. Retrowel the first area, holding the trowel at a slight angle as in step 2. With firm pressure and long strokes, sweep over the area only enough to smooth out any unevenness.

Too much retroweling may cause hairline cracks, or crazing. In addition, a slightly fibrous texture has a more pleasing appearance and hides unevenness in the surface better than does a very smooth surface.

A calcamine brush may be used to obtain a pleasing, brush surface in place of the troweling described in step 4. Brushing must be done with light strokes immediately following step 3, before the mix begins to set. In hot, dry weather brushing may need to be done on smaller areas immediately after step 2. Use either horizontal or vertical strokes depending on the surface effect desired. When the brush begins to drag because of mix collecting in the bristles, dip the brush in water and shake out the excess. This will probably have to be done after brushing an area of 10 or 15 square feet.

A stippled surface may be obtained with a paint roller from which the fibers have been burned off with a torch. The fibers melt, leaving nubs on the roller surface. Follow the same procedure as described for obtaining a brushed surface.

If the bonding applications must be stopped for 30 to 45 minutes or more, try to stop at a corner or at the edge of a window or door opening, particularly when color has been added to the mix. Color differences that might occur between batches will then be less apparent.

Fill in the corner junction between the wall and footing, carrying the bonding mix onto the top of the footing on both sides of the wall. If the wall is built on a concrete slab floor on grade, carry the surface bonding down over the outside edge of the slab to help seal the joint between wall and floor.

Wet the finished bonding with a fine spray of water once or twice the first day to aid curing . . . .

COVERAGE OF THE BONDING MIX. Twenty-five pounds (dry weight) of bonding mix should cover at least 60 square feet of wall, or about 30 square feet of wall bonded on both sides.

On the bags of surface bonding mix that we bought (Owens-Corning Fibreglas BlocBond), one-eighth of an inch application thickness is recommended. This is what we did, and the extra thickness gives me an added sense of confidence, especially in a sub-surface structure.

In place of pilasters, we filled three or four cores of blocks with concrete and rebar every 8 feet around the walls.

Waterproofing Walls

The Owens-Corning Fibreglas Corporation say in their manual, *BlocBond Construction Techniques:*

"Fibreglass Blocbond has been rated 'Excellent' for resistance to wind-driven rain at wind velocities up to 100 mph and has withstood a four-foot hydrostatic head with no water penetration. When applied properly it will prevent perceptible moisture penetration into or through even relatively permeable masonry. In areas subject to severe weather conditions, you may have to apply additional coatings or treatments. Remember, however, that cracking of foundation walls due to settlement or any other cause will nullify *any* waterproofing treatment."

Here is another reason why a heavy footing should be built on undisturbed earth: to avoid settling. As an added precaution, we applied black plastic roofing cement and 6-mil black polyethylene (overlapped at the corners) to all the walls. (This will be discussed further in the chapter on roofing.)

Remember, too, that "severe weather conditions" mentioned in the paragraph quoted above do not apply to subterranean houses.

Plates

The changeover from masonry to timber construction is accomplished by the use of wide wooden plates—or "sills"—bolted to the top of the masonry wall. With a stone wall, the anchor bolts must be built in with the last course of stones. With a poured wall, the bolts are set into the concrete while it is still workable. There are two ways of setting the anchor bolts into the cavities of blocks, and we employed both, depending on the location of the bolts. In corners and other locations where the cavities are completely filled, the bolts can be set into the concrete while it is still workable. Where an anchor bolt is required between filled core locations, stuff a wad of crumpled newspaper 18 inches into the hole and fill with concrete. Be careful when setting the anchor bolt that the paper is not pushed farther and farther down. Use a 12-inch to 16-inch anchor bolt, with the bottom of the bolt bent at right angles to the shaft to assure that the bolt cannot be pulled out once the concrete has set. Leave the bolt exposed above the top of the wall a distance equal to the thickness of the plate to be used, or just a hair less.

A good way to make sure that the bolts are plumb and at the right height is to make a template out of a piece of the plate to be used. Drill a hole through a piece of the same (or slightly larger) diameter as the bolt. Take care that this hole is exactly perpendicular to the top of the piece. This template can now be used to check both depth and plumb.

Plan the location of the anchor bolts based on the lengths of the planks you are going to use for plates. Two bolts are sufficient for short plates; use three for plates over 9 feet. Keep the bolts about 8 inches in from each end of the plate.

I recommend wooden plates at least 2 inches thick and $\frac{1}{4}$-inch wider than the width of the blocks. The exception, in our house, is on the east and west walls, where we used plates of a rather unorthodox 4-by-8 dimension. There were two reasons for this: First, the 4 inches were necessary to bring the wall up to height. Second, by keeping the plate to the inner edge of the wall, we left a surface of masonry from which we could build upwards with 4-inch-wide concrete blocks, thus keeping the outside edge of the wall flush and of masonry construction right to the top of the roof (Fig. 36).

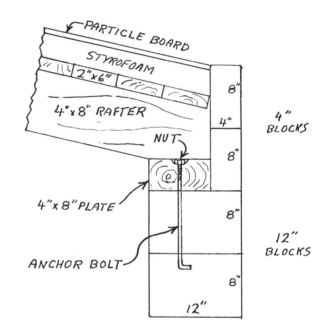

Fig. 36. Cross section of the east wall, showing the anchor bolt and the 4-by-8-inch wooden plate used to bring the wall up to height.

After the anchor bolts are permanently set in the concrete—wait a couple of days— locate the position of the holes in the plate by laying the plate on the anchor bolts and whacking once sharply with a hammer (Fig. 37). Get a friend or two to hold the plate steady and in the correct position while doing this. The indentations formed mark the location of the holes to be drilled. Turn the plate upside down and drill holes of the same diameter as the bolts, being careful to keep the brace and bit straight up and down (Fig. 38). Next, chisel out a depression to accommodate a washer and nut (Fig. 39 and 40). Then lay a thin ($\frac{1}{4}$ to $\frac{1}{2}$ inch) layer of glass fiber insulation on the top of the wall where the plate is to go, and bolt the plates down with a crescent wrench or a socket set. The glass fiber acts as a "sill sealer" and protects against drafts and insects. Glass fiber made especially for the purpose can be bought, but it is cheaper to divide a roll of regular wall insulation yourself.

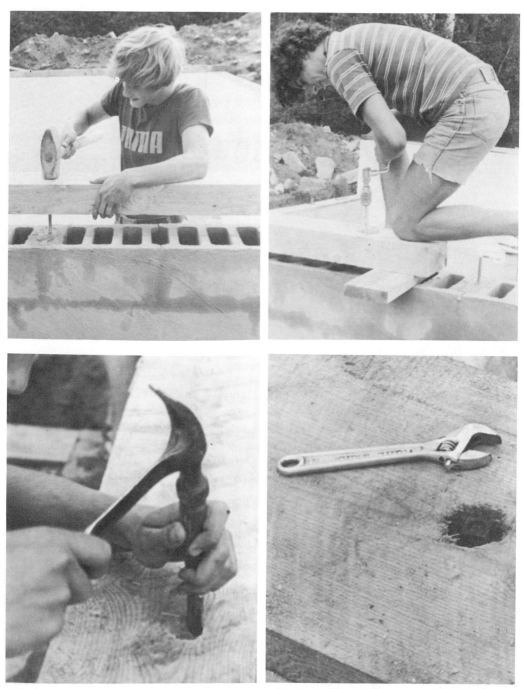

Figs. 37, 38, 39 and 40. After the anchor bolts have set in the concrete, locate the holes in the plate by laying the plate on the anchor bolts and whacking once sharply with a sledgehammer. The indentations in the plate mark the locations of the holes to be drilled. Turn the plate upside down and drill holes the same diameter as the bolts. Chisel out a depression to accommodate a washer and nut.

7. Framework

Post and Beam Framework

Old-fashioned, perhaps, in this fast-moving world of plastic foods, planned obsolescence, and stick framing with 2-by-4s that measure $1\frac{1}{2}$ inches by $3\frac{1}{2}$ inches, but post and beam framing is still recognized as one of the strongest ways to build. It is my favorite framing technique, not only because it is strong, but because one can feel the strength in an almost ethereal way, especially in structures where the framing is left exposed. The same holds true for plank and beam roofiing, the natural bedfellow of post and beam framing. Most people like the rustic atmosphere of old inns in Europe, England, and New England, with their exposed, hand-hewn timbers enclosing white plaster panels. I think the folks who enjoy that kind of atmosphere are probably picking up a sense of security, subconsciously perhaps, from seeing those monolithic beams overhead. You know they aren't going to fall in on you. All over England, thousands of ancient buildings still stand to "bare" witness to the structural advantages of post and beam framing.

You can have timbers milled from selected logs, or you can use old hand-hewn barn timbers. I chose the barn timbers for the post and beam framework and had the roof rafters and 2-by-6 planking milled from straight hemlocks, which I selected myself while they were still standing. There were three reasons why I decided on barn timbers:

One, they are incredibly strong. Most of the main support beams used in New York and New England during the nineteenth century are hewn to 8-by-8 or 10-by-10 dimensions. What's more, the trees were finer-grained in the old days, before planned forestation was introduced to encourage rapid growth. Of course, any beams that are rotten, whether from insects or moisture, should be immediately rejected. Choose carefully, and store any wood up off the ground while it is waiting to be used. With barn beams, stack them so that water cannot collect in the hollows (mortises) that were chiseled out for mortise and tenon joints.

Two, old barn timbers are no longer green and have finished shrinking. This may not make a lot of difference with most kinds of construction, but our intention was to employ the "stovewood masonry" (log-end) style of building in many of the panels framed by the posts and beams. Green wood will shrink away from masonry infilling.

Three, we like the character of the old beams—the mortises, the adze marks, the dowel holes. We feel that the wood's character is imparted into the atmosphere of the finished room. Romantic? Maybe. But it's all part of what gives the home a high L.Q. (Livability Quotient).

We looked for suitable rafters among old barns and houses that were being torn down in our area and found a few that would have done the job, but not enough of consistent size, so we decided to go with new 4-by-8 hemlock rafters. Shrinkage on exposed rafters doesn't really effect the construction in any way and, as we were able to choose our own straight hemlocks, we knew the timbers would be strong. We also bought the wood for the planking from the same supplier.

Green wood should be stacked on site several months before you plan on using it, to lessen shrinkage and the chance of rot. As I say, shrinkage on the rafters doesn't matter too much, but you don't want $\frac{3}{8}$-inch spaces between the roofing planks. The ideal roof planking would be seasoned tongue-and-grooved 2-by-6s, but these are expensive. If you buy them, make darned sure they're properly dried so that the tongues won't pop out of the grooves. If in doubt, take a sample home, cook it in a 400-degree oven for a few hours, and see how much it has shrunk. If you're lucky enough to score an old silo that's being pulled down, as we did while building the cottage, you've got the perfect roof planking: old, dry, full-sized, tongue-and-grooved, 2-by-6 planking. Spruce, if you're really lucky. Ask around. Some farmer may be willing to sell a leaning silo that he's not using anymore. We bought ours for $100. You can figure the board feet and make him an offer. And don't forget: those old silo hoops make great rebar for footings and block wall cores, so don't leave them behind.

A big asset to building economically is being a good scrounger. And old recycled materials are often superior to stuff being produced today.

We built Log-End Cave around the three major support members, 30-foot 10-by-10 barn beams, which we bought for a dollar a running foot. They are beautiful timbers, exposed throughout the house, a real design feature of the construction. We tried to plan our post location to take into account both structural considerations and practicality with regard to the floor plan. We chose the biggest of the three timbers as the center support beam. I did not want to span more than 10 feet with an old barn beam, even though it was in excellent conditon and was a full-sized spruce 10-by-10, so I planned for two posts, one out of the way on the stone hearth, where it would help to support firewood, and the other between the kitchen and dining areas. All three spans thus created are less than 10 feet.

Since the beam had a slight taper—from 10-by-11 inches to 10-by-9 inches—we placed the "weak" end of the beam to the south, where the first span is only 7 feet 3 inches. The floor plan was designed so that the north-south internal walls would fall under the other two 30-footers, so they are supported by three major posts each, dividing the depth of the house into approximate quarters. We placed the posts at the intersections of the internal walls, so that the 4-inch walls could butt against the 8-by-8 posts, leaving them exposed in the corners of the peripheral rooms. The two peripheral 30-footers are positioned exactly between the center support beam and the east and west side walls, so that the spans of the 18-foot rafters are consistently 8 feet. Another advantage of this plan is that the long north-south internal walls rise to meet the underside of the long beams, so there is no tricky fitting of the walls around the exposed rafters. The only exception to this is where Rohan's room juts out into the living room area, necessary to give him a reasonably sized room. How the internal wall was fitted around the rafters can be seen in Figure 41.

Fig. 41. One room juts out into the living room area, the only place where the walls had to be specially fitted around the exposed rafters.

While on the subject of the integration of the floor plan with the support structure, I should mention that all the east-west internal walls rise up to meet the underside of a 4-by-8, except for the wall between the office and the master bedroom, which meets the ceiling halfway between two rafters. There are two reasons for this: one, it saves slightly on materials, because the 8 inches of the rafter are used as part of the internal wall, and, two, it looks a lot better than just missing a rafter with a wall (which would also give a convenient place for cobwebs to start). Because of space considerations, we varied slightly from this plan with the office-bedroom wall, splitting the difference between rafters in this case.

Setting Up the Posts

The building plan called for a 22-inch space between the top of the block wall and the underside of the center beam. Subtracting 2 inches for the plate, we were left with a 20-inch post to support the north end of the beam. We called the floor level and based the heights of the other center posts on the total height of the beam off the floor at the north end, 95 inches (7 feet 11 inches). We cut the two center posts exactly to 8 feet so that we could check them an inch into the beam for rigidity.

Since the south-wall post was in two pieces, it was a bit trickier to figure. The base post, the one standing on the wooden plate, had to be 48½ inches to accommodate the 4-foot-high windows we'd planned for the south wall. The header for the windows was to be a 4-by-10 laid on its side, 4 inches for resistance against sagging, 10 inches to establish the width of the wall for the log-ends later on. The short post on the top, then, was all that was left of the 95 inches of height established at the north end. The subtraction was: 95 inches minus 28½ inches (the block wall) minus 2 inches (the plate) minus 48½ inches (the base post) minus 4 inches (the header), leaving a short "post" of 12 inches. A mason's line stretched from the top of the north wall post to the south wall post showed one of the center posts sticking up 1 inch, the other 7/8 inch. The floor was pretty level after all.

Having removed all nails and marked the rough hewn posts as best we could with a square, we cut one end with a chainsaw and stood the post upright to check for plumb. If the post tilted in any direction, it was necessary to recut the bottom, compensating for the error. Only when the bottom of the post was flat did we then measure for the cut at the top. You can use the square again on the top measurement, but I've found that it's less risky to measure the length of the post from the good end along each of the four corners and then connect the marks. The irregularities of a barn beam can throw the square off by quite a bit, and there's no room for error on the second cut.

It is not advisable to stand wood directly on a concrete floor, due to the possibility of trapping dampness, so we cut a square of 50-pound roll roofing to stand each post on. The roll roofing also seems to steady the post somewhat. We waited until all eight internal posts were cut and squared before standing them up permanently, in case one got knocked over and injured someone. Calculating the heights of posts that support the two peripheral beams is a little tricky and can best be explained with the aid of a diagram (Fig. 42).

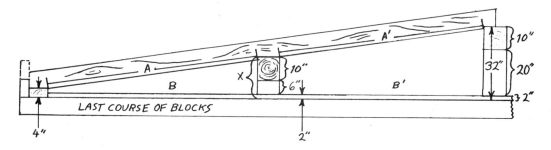

Fig. 42. Calculating the heights of the internal posts that will support the two peripheral beams.

To make the best structural use of the 4-by-8 rafters, I wanted to make the spans A and A' equidistant. This will be accomplished if the post location is established on the plate such that B = B'. Now consider the total height of the supports for the 4-by-8 as measured from the top of the block wall. At the lower end, the rafter is to be supported directly by the 4-inch-high plate. The high end of the rafter is supported by a 2-inch plate, a 20-inch post, and a 10-inch beam—32 inches in all. The height of X, then, should be midway between the 4-inch and the 32-inch figures, or 18 inches. Subtracting 2 inches for the plate and 10 inches for the support beam leaves a 6-inch post. I cannot recommend actually using a "post" as short as 6 inches, because a heavy load might cause it to split open, just as a thin slab of wood is easier to split with an axe than is a thick one. Instead, I used two short pieces cut from a scrap of old 3-by-10, stacked

67

one upon the other. These worked perfectly: they were exactly the right dimensions and were easy to fasten to the plate with 20-penny nails. A measurement from the floor established the height of the internal posts, as discussed previously for the center beam.

The support structure of our house is actually a very simple one and I hope that by examining in some detail the ways in which we met our problems, the reader will gain a sense of the principles involved, enabling him to solve the structural problems of a somewhat different design that he might wish to employ. The mathematics involved is never more than simple arithmetic and the most basic geometry. Keep a set of plans nearby, make notes on them based on actual (not theoretical) measurements, and refer to them constantly, double-checking your figures from two different approaches, if possible.

When all eight internal posts were squared and cut, we stood them up on their squares of roll roofing at the correct premeasured locations, shimming if, necessary, to assure perfect plumb. We made them rigid with a rough framework of 2-by-4s (Fig. 43) We did not attempt to fasten the posts to the concrete floor. The tons of weight that these posts support, as well as the internal wall tie-ins, assure that the posts will not move once the roof is on. The rough framework's purpose is to hold the posts upright while putting the massive 10-by-10s in place. We did not dismantle this temporary framework until after the rafters were fixed.

Fig. 43. After the internal posts are squared and cut, stand them up on their squares of roll roofing at the premeasured locations and shim, if necessary, to assure perfect plumb. The internal posts can be kept in place with a rough framework of 2-by-4s.

The window and door header (or lintel) on the south side of the house had to be put into place before the small post for the south end of the center beam could be permanently located. This lintel consists of a pair of 17-foot 4-by-10s, cut for the purpose at our local sawmill. It took four of us—Steve, Bruce, Jaki, and myself—to maneuver the things into place. The south ends of the other 30-footers would rest directly on the headers at the points of their greatest strength: right over stout 8-by-8 barn posts (see south-wall elevational plan, Fig. 3).

The plank and beam roof of our house is carried by the east and west block walls and the three aforementioned 10-by-10s. In reality, they vary quite a bit from beam to beam and from end to end, but the smallest dimension at any of the six ends is a full 8-by-9 inches, and the average size is very close to a full 10-by-10.

The beams were stacked next door, almost a quarter mile from the cave site. Steve, Bruce, and I took very careful measurements of the locations of the posts and transposed the measurements onto the actual timbers. We'd chosen the heaviest of the three beams as the center beam, and labeled the others "East" and "West."

We squared the ends of the beams. The total length of the center beam is 30 feet 3 inches, which allows the ends to extend 4 inches beyond the support posts for the sake of appearance. The west beam was cut at 29 feet 11 inches to give a 2-inch over-hang at each end. We wanted to make the east beam the same, but we could only get a good 29 feet 8 inches out of it. We solved the problem later by adding a 3-inch piece from another beam, which in no way affected the structural considerations because the last rafter still had good solid wood to rest upon. This cosmetic surgery is only noticeable if it is pointed out, as all the external surfaces of the beams were later creosoted to a uniform color.

I went out to recruit some able bodies while the boys finished up the checks in the beams to accommodate the internal posts. As usual, it was a hectic day of running around making last minute preparations. By late afternoon, I'd gathered four strong bodies to supplement Steve and Bruce. We backed the pick-up to the western beam, the lightest of the three. We figured that if we couldn't move that one, we'd have no chance with the 600-pound center beam. With Jaki looking after the baby, I was left with the easy job: driving the truck. Six men lifted the north end of the timber while I backed under. The beam was now tucked right up to the back of the cab. Then, paired at each end of strong ash branches, the three teams lifted the south end off the ground and shouted the command to "Go!" Like some great crawling insect, we started up the hill. Once, halfway up the hill, I was chastised by six screaming banshees, in no uncertain terms, to quit speeding. We purposely overshot the driveway and backed up to the house site. We backed the beam so that almost half of it was cantilevered over the edge of the north wall, and then manhandled it from below into position on the posts (Fig. 44). To our great delight, the post fit perfectly into the checks. So much for the easy one. The boys voted to tackle the biggie next, reasoning not unwisely that they mightn't be able to manage it after hauling another of the small 500-pounders.

Not only was the center beam considerably heavier than the other two, but it had to be raised almost 2 feet higher. We manhandled it halfway out over the hole, as before, and then put our strongest and tallest man, Ron Light, below the beam, catwalking with Bruce along some makeshift scaffolding. Almost single-handedly, like Big Bad John of the song, Ron passed the south end up to where Paul and I could get our ash branch

Fig. 44. The beam was then cantilevered over the north wall and positioned on the posts.

under it. Then, with a monumental effort by all, we hauled it up and into place. Again, the checks were beautifully positioned but, alas, one was not deep enough. "It's gotta come off," I said. Groans from the crew. "Maybe the post will compress," suggested one of the tired backs hopefully. We took the beam off the posts and laid it to one side while I removed 3/8 of an inch from the top of a post with my chainsaw. Steve did a little chisel work on the check. Back went the beam, this time fitting almost perfectly. We'd actually overcompensated slightly and the beam did not rest on the post by an eighth of an inch. Good enough. We knew the beam would sag an eighth under the roof load and rest on the post as intended.

The last beam, by comparison, was a cakewalk. Easy for me to say. I was still driving the truck!

Plank and Beam Roofing

Our 4-by-8 rafters are the "beams" of plank and beam construction. Why 4-by-8s? Several reasons. A 4-by-8 is proof against twisting under a weight load. The base of the rafter is so wide that it sits unassisted on the support beams. Toenails are sufficient to keep it in place. With 2-bys, it is necessary to cross-brace adjacent rafters with metal ties or wooden blocks to protect against twisting under a weight load. The 8-inch measurement provides excellent strength on the relatively short 8-foot spans and has the advantage of being twice the 4-inch measurement. This makes for economic use of the log and gives a well-proportioned rafter, pleasing to the eye if left exposed. The alternative size which we considered was a 3-by-10 rafter, as we'd used at the cottage. These, too, fit nicely into the cross section of a log, but they require larger trees, which may be difficult to find nowadays. The reader will readily see in Figure 45 that measurements such as 3-by-8 and 4-by-10 do not return the maximum number of rafters from round logs.

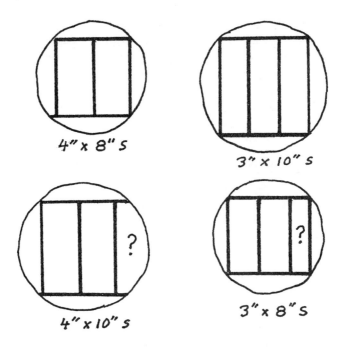

4" x 8" S

3" x 10" S

?

4" x 10" s

?

3" x 8" S

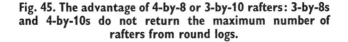

HEAVY LINES INDICATE SAWCUTS

Fig. 45. The advantage of 4-by-8 or 3-by-10 rafters: 3-by-8s and 4-by-10s do not return the maximum number of rafters from round logs.

We used hemlock for its inherent strength as a rafter. Spruce would have been even stronger, but decent-sized spruce is hard to come by in our area. The reader should use the strongest wood grown locally, but increase dimensions or decrease spans with woods not as strong as hemlock.

We put our rafters on 32-inch centers mainly for strength. We tailored our floor plan and the east and west wall lengths to multiples of the 32-inch centers. (This also worked in well with the blocks.) Someone accustomed to stick-framing and plywood may think that 32-inch centers are rather wide and, for that type of building, they would be. But plank and beam construction permits up to 8-foot centers to carry a "normal" roof. A sod roof, of course, is not normal, so we went with 32-inch centers. In combination with 2-by-6 hemlock planking, our roof will support the 120 pounds per square foot which could be expected in the worst conditions, such as when a long, hard, warm spring rain goes to work on 4 feet of snow.

Given 2-by-6 planking as a constant, there is one other strength consideration in addition to rafter dimensions and frequency, and that is span. In our house, rafters always span exactly 8 feet, never more. This is not asking much from a 4-by-8 hemlock, even given a sod roof and a snow load. Increasing this strength is the fact that each 18-foot rafter is supported in the middle by an unshakeable 10-by-10 beam. One rafter covering two 8-foot spans is much stronger than two short rafters each spanning 8 feet, due to the cantilevering effect of one load offsetting the other (Fig. 46).

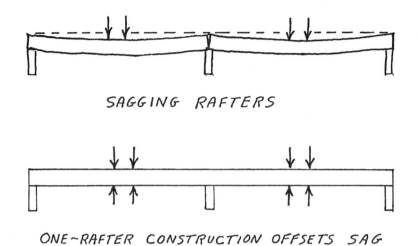

SAGGING RAFTERS

ONE-RAFTER CONSTRUCTION OFFSETS SAG

Fig. 46. One rafter covering two 8-foot spans offsets sag and is much stronger than two rafters spanning 8 feet each.

Installing the Rafters

The three big beams were in place, but the rafters were still in the form of logs. At least they were at the sawmill. The boys and I had had to spend two days at the property where I bought the hemlock to speed up progress and get the wood out. The order was already seven weeks overdue. Bad enough that it cost us time on the job, but it meant that we would be putting the plank and beam roof up green, causing considerable shrinkage.

When dealing with rafters with birdsmouths (checks designed to enable the rafter to sit flat on the beam), the fastest and easiest method is to first make a template (Fig. 47).

The template will seldom fit exactly right if you are using old barn timbers, but it will always fit close enough so that you can make a note of any necessary corrections when marking the actual rafter. For example, where a beam is high, you will have to make the birdsmouth deeper; if low, the birdsmouth will be smaller. You will devise your own method of making corrections, but the template will remain valuable because the birdsmouths cut into it are are always the right shape and it can be moved as necessary to make corrections.

Steve and Bruce would take readings with the template. Bruce might say, "The center birdsmouth needs to be a quarter-inch closer to the peak." They bring the template over to where we're cutting and we mark it as required (Fig. 48). While I cut, they go back and get the reading for the next rafter. When a rafter is ready to be put in place, we use three men for the job. Two can slide it along the beams as shown in Figure 49. After snapping the picture, I moved into position on the center beam so the guys could slide the rafter up to me. Of course, it fits perfectly . . . most of the time. Figure 50 shows half of the rafters in place.

Do not nail the rafters into place as you go. Rather, wait until all the rafters are in place, level them, and then toenail them into the plates and beams. One way of levelling the rafters is to stretch a mason's line from the first to the last one, shimmed a half-inch off the top of these so that the line does not actually touch any of the in-between rafters.

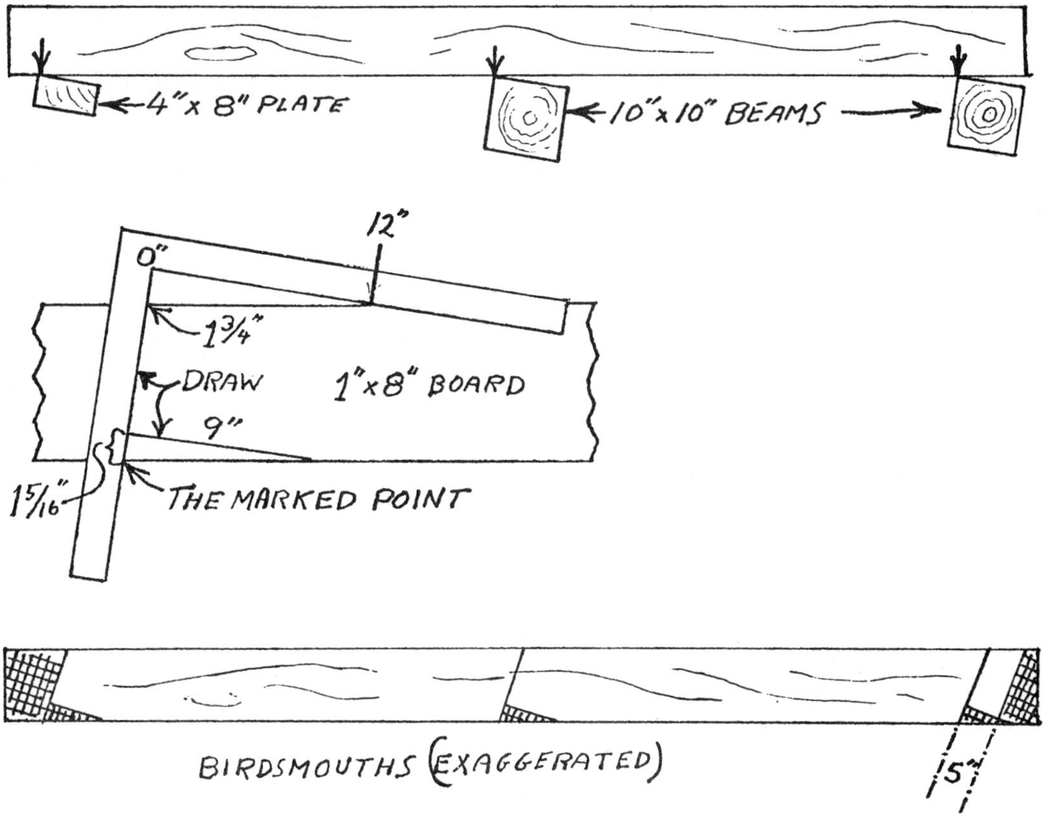

4"x 8" PLATE ← 10"x 10" BEAMS →

12"

0"

1¾"

DRAW   1"x 8" BOARD

9"

1⁵⁄₁₆"   THE MARKED POINT

BIRDSMOUTHS (EXAGGERATED)   5"

Set a long board on the roof (we used a 1-by-8) and mark the location of the three corners of the supporting members. Take the board down. Using a carpenter's square to determine the correct pitch, draw an extension of the edges of the supporting beams. Our pitch, for example, is 1.75:12. We want about 9 inches of each rafter to sit flat on each of the supports. As 9 inches is $\frac{3}{4}$ of 12 inches, the depth of the birdsmouth will be $\frac{3}{4}$ of 1$\frac{3}{4}$ inches, or 1$\frac{5}{16}$ inches.. Using the square again, draw a line perpendicular to the first lines drawn, beginning at a point 1$\frac{5}{16}$ inches along the first lines. The little triangles are the required birdsmouths. Remove the excess wood as shown by the crosshatching. Note that another perpendicular line rises at a point five inches along the topmost birdsmouth, so that the rafters will butt over the center beam.

Fig. 47. Template for cutting birdsmouths into the rafters.

Fig. 48. Using a template will ensure correct marking for the birdsmouths on each rafter.

Fig. 49. The rafter is slid along the beams and put into position.

Fig. 50. Half of the rafters in place.

It will be easy to measure for rafters that need to be shimmed or checked out in order to fall in line with their companions. Make sure that neither of the rafters you are using for guides are way out of line themselves. If so, correct the culprit right away and proceed. How fussy to be? Well, that's up to the individual. On 32-inch centers, though, an eighth of an inch variance is not too serious.

We Interrupt This Story to . . .

If this were purely an instructional manual on how to build a particular house, it would be unforgivable to stop halfway through a discussion of plank and beam roofing for any reason whatsoever. But it is not simply an instructional manual; it is also the story of an owner-builder wrestling with a little-known style of building. The reality is that we were interrupted by unforeseen circumstances, and the reader might as well expect that he, too, will be confronted with such circumstances, though I offer him my best wishes that his will be not quite as unforeseen as ours.

The completion of the rafters on August 26 marked the end of a productive period that began on July 5 when the digger took its first big bite out of the knoll. Except for the delay in getting the hemlock to the sawmill, progress had been swift, at least from an owner-builder's point of view. Jon Cross, my contractor friend, might have considered our pace to be that of a particularly sluggish snail. Still, fifty-three days after the start of construction, we were ready to put the roof on . . . if the roof had been ready for us. And if I hadn't fallen off a rafter onto the window header below, badly bruising, if not cracking, a rib. Stupidly, I was catwalking an unnailed rafter to get to the peak to help with the next rafter. Our rafters are the same width as a gymnast's balance beam, but I lack the agility of a Nadia Comaneci, who might have recovered her balance when the rafter fell on its side and off the south end of the house. I did a spectacular dismount, which was interrupted before its conclusion by an unshakeable 4-by-10. Still, that was probably preferable to hitting the ground a fraction of a second before the 180-pound rafter.

Fifteen minutes later, I was back on the beams, a wiser man. We finished the rafters, and I laid blocks the next day, but by four or five days after the tumble, I could not do any physical work, and the project pretty much came to a halt for a week. I used the time to try to solve the next problem, which was that we were about 15 percent short of the required roof planking. There was no way I was going back to my original supplier. I couldn't wait another six weeks for the wood. I was forced to buy five plantation pines and haul them to the sawmill. I decided that as long as I mixed the pine planks carefully with the hemlock, I would not be sacrificing too much in the way of strength. I think this turned out to be a correct assumption, though the pine was very green indeed and began to go a little moldy in the interior until I connected the stove and dried it out. Today, the effect is that every fifth or sixth board has a kind of a bluish-gray color, rather nice, really, though not exactly what we planned on. Over the sauna we used hemlock exclusively, fearing that the 200-degree temperature might cause the pine pitch to bleed from the ceiling. Eastern hemlock is not a pitchy wood.

The concrete blocks I laid were the first course of 4-inch blocks to go on top of the 12-inch block wall. The reader will recall that our east and west plates were only 8 inches wide in order to leave room for these blocks (Fig. 36). I did not use surface bonding because it could only have been applied to the wall's exterior. (It may occur to the reader that surface bonding could be used on the exterior of the 4-inch blocks as a waterproofing. As I did not have sufficient surface bonding material left to do this, I decided to wait and waterproof later with black plastic roofing cement and 6-mil black polyethylene, as described later.) This portion of the wall was not loadbearing, so I felt confident in laying up the blocks with mortar. I decided to lay the first course before nailing the roof planking, so that the planking would not be in the way of my work, but I held off on the second course so that the blocks would not be in the way of the planking.

Here I should mention a problem the reader can easily avoid. We'd spread about half of the planks on the roof, loose, to give them a good place to dry. Before any rain, we laid out rolls of 15-pound felt paper, lapped, to try to keep as much water out of the house as possible. We could not start nailing the roof, as we were still waiting for the pine logs to be milled into planking. One night, the first strip of felt paper blew up and backfolded onto the roof. Water poured down the roof and onto the top of the 4-inch blocks, finding its way finally into the unfilled cores of the 12-inch blocks. There is no way to get that water out short of drilling holes into the block cavities (from the inside, so as not to break the waterproofing quality of the surface bonding). This we did. The moral: maintain vigilance in keeping water out of the wall cavities.

While I was laying up the first course of 4-inch blocks, Steve and Bruce started troweling black plastic roofing cement onto the north wall. Later in the day, we pressed sheets of 2-inch Styrofoam onto the wall, hoping that all that plastic cement would hold it in place. It did. For a while.

Steve and Bruce left on August 28. Their help had been invaluable. I think my diary best captures the mood of the next twelve days:

"Monday, August 29. Hauled 2-by-6 planks from the sawmill."

Progress was so slow that the next eleven days are all lumped together:

"Tuesday, August 30—Friday, September 9. Discover that we will run short of planking (thought we had more than enough), so I chase around for hemlock. No luck, so I work a deal with Dennis for pine. What we think is Norway pine turns out to be Plantation pine. Gary and Diane arrive on Labor Day weekend and Gary helps Dennis and me clear a road to the pine. I can't do much because of my cracked rib. Sunday, Gary babysits while Jaki and I help Dennis. Following week, we get wood to sawmill for milling. Dennis and I start roof with hemlock during this time and cut three or four trees in front of cave. Get five hemlock and six spruce 2-by-6s from our own trees, which had to be cleared to let the winter sun get to the south wall. Depressing time, during which little gets done."

So it goes. On September 10, we helped the sawyer plane the pine, spruce, and hemlock 2-by-6s to uniform thickness and width. This is important in getting a good tight fit when roofing. In addition the planed surface is a lot less prone to collecting cobwebs. A little wood (and strength) is lost, of course, but not enough to call it significant. Our planking actually measures $1\frac{3}{4}$ by $5\frac{3}{4}$ inches.

8. The Roof

In planking, start at the bottom and work up. Get the first course of planking dead straight and make sure each end of the first course is equidistant from the peak. Snap a chalkline to mark the bottom edge. We had several long planks that were cut from the same logs as the rafters, so we were able to make the first course on each side of the roof from two planks each—a good start.

A word about lengths of planks and wastage: There are twelve rafters or eleven "spans" on each side of our house. A course might consist of two planks which will cover four spans and one which will cover three spans, for example. Each span is 32 inches, or 2 feet 8 inches. However, the planks at the ends of each course must be long enough to accommodate an overhang. I figure an extra 12 inches for this, 10 for the overhang and 2 to get to the outside edge of the outermost rafters. (The 32-inch figure is measured center to center.) To save on wastage, have the saw logs cut to useful lengths (Fig. 51). Take a copy of these lengths with you into the woods, or supply the man cutting for you with a list. Make sure that the logs are cut 2 or 3 inches *longer* than these measurements, to allow for squaring the ends of the planks.

If you are stuck with green wood, as we were, nail the planks as shown in Figures 52 and 53. There is no reason why you can't draw two boards at a time. First, start the nails as shown. Then, take a star chisel or a 12-inch spike and whack it about 3/8 inch into the rafter tight against the plank. Using the spike as a lever, draw the boards tight against each other and nail the lower nails first, then the upper ones. The reason for angling the nails is to make it easier for the wood to shrink without splitting (Fig. 54).

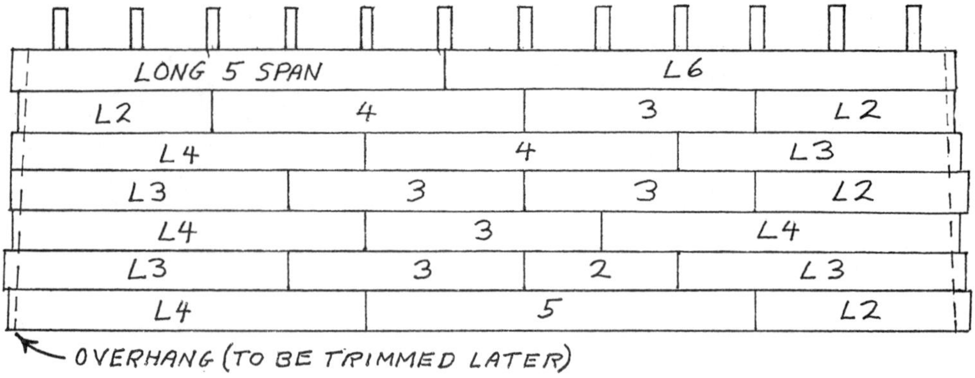

LONG 5 SPAN		L6	
L2	4	3	L2
L4	4		L3
L3	3	3	L2
L4	3		L4
L3	3	2	L3
L4	5		L2

← OVERHANG (TO BE TRIMMED LATER)

SPANS	LENGTH	SPANS	LENGTH
2	5'4"*	4	10'8"
LONG 2	6'4"	LONG 4	11'8"
3	8'	5	13'4"
LONG 3	9'	LONG 5	14'4"

* TOO SHORT FOR THE SAWMILL; CUT FOURS IN HALF

Fig. 51. Save wastage by cutting the logs to useful lengths.

Figs. 52 and 53. If you are using green wood, nail the planks as shown here, using a spike to keep them tight against each other and nailing the lower nails first.

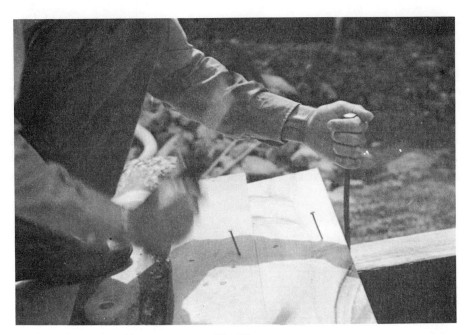

Fig. 53.

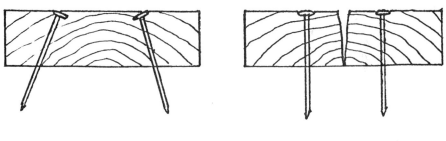

NAILS "GIVE"   WOOD SPLITS

Fig. 54. Angling the nails makes it easier for the wood to shrink without splitting.

Incidentally, I like to use 16-penny resin-coated nails, though uncoated 16s would probably do. Resin-coated nails are hell to take out once they're in, but that's the idea of them. And the narrower shaft makes splitting at the ends of the plank less likely and makes the plank more likely to give with shrinkage.

Try to avoid placing too many joins one above the other (Fig. 55). Leave the planks sticking out over the ends. The overhang can be trimmed later by snapping a chalkline and cutting the whole edge at once with a circular saw or a chainsaw (Fig. 56). I found it much easier going with the chainsaw. Green 2-bys will choke all but the toughest circular saws. The only problem with the chainsaw is that you've got to be very careful to keep it perpendicular, assuring a straight edge.

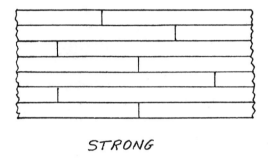

STRONG

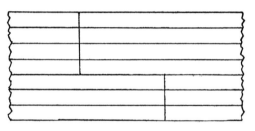

NOT SO STRONG

Fig. 55. Avoid placing too many joins one above the other.

Three things can happen at the peak (Fig 57). One, the bottom sides of the opposing plank roofs may meet perfectly. In this unlikely case, crack open a beer and relax. Two, they will just miss meeting each other. This is what happened to us. In this case, make a wedge from leftover pieces and fit it into the space. You have to use a variable-angle table saw or good skill saw for this. Three, the last course on each side will be just too wide to fit. In this case, each of the last courses must be trimmed to the correct width. It's up to you if you want to incorporate the proper angle into the trim cut; it's not entirely necessary, as there is still lots of Styrofoam and particle board to be put on.

The fascia boards—or retaining boards—serve to keep the sod roof from washing out over the gable ends of the house. They frame the roof with high walls, directing all drainage down the slope of the roof. We used 2-by-12s, which seemed to have worked out well. They should be installed after the overhang is trimmed. Our method of

Fig. 56. Leave the plank ends sticking out; the overhang can be trimmed later by snapping a chalkline and cutting the whole edge at once.

PLANKS

PERFECT FIT

TRIM TOP OF RAFTERS IF NECESSARY

ANGLE CUT OPTIONAL

Fig. 57. At the peak, the planks might fit perfectly or just miss each other. Or the last course on each side may be too wide.

fastening the fascia board involved first nailing a 2-by-6 plank right along the edge of the roof, for its full length, then nailing scraps of economy-grade 2-by-4s on top of the plank (Fig. 58). Together, the planks and scraps were now $3\frac{1}{4}$ inches higher than the roofing planks, exactly the thickness of our Styrofoam. (If 3 inches of Styrofoam are to be used, a double course of economy 2-by-4s—actually $1\frac{1}{2}$-by-$3\frac{1}{2}$ inches—would do the same job.) This made it much easier to nail on the particle board later. The planks and scraps provided a high edge for nailing the fascia board. (Don't forget to creosote the fascia board before nailing it up.) After the fascia board was well nailed, $\frac{5}{16}$-by-4-inch lag screws were used to make sure it was on permanently—a lag screw every 3 feet or so along the fascia.

It is a good idea to use a chainsaw to cut a groove $\frac{1}{4}$-inch-wide by $\frac{1}{4}$-inch-deep down the middle of the bottom side of the fascia board before nailing it in place (Fig. 59). This groove serves as a "drip edge" and prevents rainwater from running along the underside of the roof planking and into the house.

Fig. 58. Fascia boards were fastened by first nailing a 2-by-6 plank along the edge of the roof, then nailing scraps of economy-grade 2-by-4s on top of the plank.

Fig. 59. Use a chainsaw to cut a groove down the middle of the fascia board before nailing it into place. This groove will serve as a "drip edge" to keep rainwater from running along the underside of the roof planking and into the house.

Styrofoam Insulation and Particle Board

If you can't get to the insulation right away and want to avoid bailing after a rain, put down a temporary cover. We used 15-pound felt paper for this purpose because we knew we'd be making permanent use of it later anyway. You can also use it as you progress up the roof with your planking to keep at least some rain out during that job. Before the roof is on, water is not much of a problem—evaporation will pick up what can't be pushed out the front door with a half sheet of plywood. But once the roof is on, evaporation is not nearly as effective an ally. We weighted the paper down with scrap boards to keep it from blowing away, successive sheets overlapping the previous sheet by a couple of inches.

Now is the time to frame for skylights. Follow the manufacturer's recommendations for sizing and have the actual skylight handy to check against. Remember to box the skylights in high enough to allow for the Styrofoam, the insulation board, and the sod. I found that the easiest way to box them in is to cut out the opening in the roof with a chainsaw, build the box frame on a flat surface, and then nail the box frame around the opening (Fig. 60). Our box frames are constructed of a double course of 2-by-6 planks, but a single course of 2-by-12s would be easier. Stovepipe holes can be cut later.

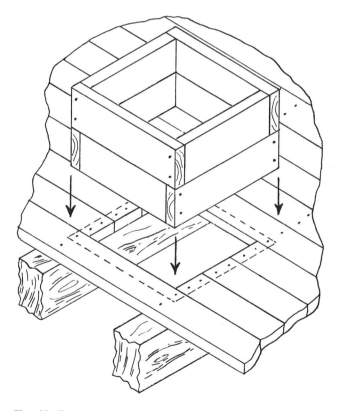

Fig. 60. To accommodate skylights in the roof, build the box frame on a flat surface and then nail it around the opening.

Styrofoam is one of the key materials in subterranean construction. It is virtually bio-*un*degradable and excellent for use on the exterior of the block walls. It is the most sensible insulation for the roof because it obviates the need for a "double" roof.

Styrofoam should not be confused with polyurethane foam, which has a higher insulative value (R8 vs. R5. The higher the "R" value, the greater the resistance to heat loss), but gives off a deadly poisonous gas if ignited. It is true that the likelihood of the insulation ever catching fire over a plank and beam roof is slight, but I just don't like the idea of it. Besides, polyurethane foam is much more expensive than Styrofoam.

Styrofoam comes in many different shapes and sizes; 4-by-8-foot sheets are among the most practical. The small 16-by-48-inch sheets are outrageously expensive. We were lucky in obtaining all the Styrofoam for our roof for $105. We bought two stacks of used Styrofoam which had been salvaged by a couple of roofing contractors in our area. These sheets were of Canadian make and measured 2 feet by 4 feet by $3\frac{1}{4}$ inches thick. The most unusual feature of these sheets was that they were shiplapped in both directions, so that all joints were covered when they were laid out on the roof. This feature further decreases heat loss. I was told that new, these sheets would cost about $5 each, so I guess we saved about $500 by buying them used. Of course, we probably wouldn't have bought such fancy stuff if we had been buying new roofing insulation. I think I would have gone with 2-inch thick 4-by-8 sheets, which we would have lapped in the other direction with 1-inch sheets. This would come out to about half the cost of the fancy shiplapped material new.

(A hint about Styrofoam prices for readers living near Canada: My experience over the last three years has been that Canadian Styrofoam is generally about half the cost of American Styrofoam. And, for us in New York, the savings on sales tax completely offsets the customs duty. You can also ask around among your local roofing contractors. You may very well get a bargain on sheets that have been salvaged from buildings due for demolition or reroofing. They might be dirty or chipped, but these aren't really drawbacks as the sheets will be covered anyway. And you might save hundreds of dollars.)

How much Styrofoam to use? In northern climes, use at least 3 inches, 4 if you can figure out a way to fasten it to the roof without railway spikes. I couldn't, but I already had $3\frac{1}{4}$-inch sheets anyway. Three inches of Styrofoam may not seem like much insulation in these times of energy shortage, when the government is recommending 10 inches of fiberglass on the roofs of new houses. But don't forget that the hemlock adds R2, the sod is good for R4, and the snow that accumulates on this type of roof is of tremendous insulative value. Remember, too, that there is no heat loss through the rafters as with ordinary roof framing. When all these factors are considered, I would match our roof's ability to hold heat in winter with the best of the rest.

Applying the Styrofoam to the roof is one of the easiest and least time-consuming steps in the whole construction. If the air is calm, it isn't even necessary to nail the sheets, as the particle board will be nailed anyway, holding the Styrofoam sandwiched between. You can progress with both jobs together or lay all the Styrofoam one day and the particle board the next. In the latter case, hammer in a single large nail with a suitable plastic or metal washer to prevent the 2-by-4-foot panels from blowing away, two or three nails on 4-by-8-foot sheets. Cover the Styrofoam for the night.

Styrofoam cuts easily with a handsaw, so it is a simple matter to trim the panels to fit around the skylights. There is no advantage to insulating the overhang, except perhaps to provide a flat surface as a base for the particle board. As mentioned earlier, we used the nailers along the edge of the fascia board for that purpose.

The particle board (also known as chipboard) provides a hard, flat surface for the application of the waterproof membrane. It serves no other purpose, although I suppose it might add about R$\frac{1}{2}$ to the insulative value of the roof. I used $\frac{5}{8}$-inch particle board, mainly because I got a good deal on it when it came on sale, but I imagine that $\frac{1}{2}$-inch board would do just as well. Plywood would do too, of course, but it is generally a lot more expensive than particle board, unless it is salvage. I fastened the 4-by-8-foot sheets to the planking with 5-inch spikes, six to a sheet. (Drill nailholes through the particle board first or the sheets will break into pieces.) I tried to keep the spikes about 4 inches in from the edge of the sheets. Near the fascia boards, 10-penny nails will suffice. In Figure 61, I am fitting a sheet of particle board up to the fascia.

Fig. 61. Fitting a sheet of particle board up to the fascia.

Now is the time to cut the holes for the stovepipe. Take your location measurements carefully from the inside and transpose the measurements to the roof. We used Metalbestos stovepipe, which is tested safe to within 2 inches of combustible material. Therefore, we cut a hole with a diameter of 4 inches more than that of the stovepipe. The only tool for the job is a chainsaw, but take care to hold the saw extra tightly while cutting through the particle board with the tip. Figure 62 shows the planking, the Styrofoam, the particle board, and the first strip of felt paper.

Before discussing the waterproof membrane, I must advise the reader of an alternative approach to the application of Styrofoam as described in *Earth Sheltered Housing*

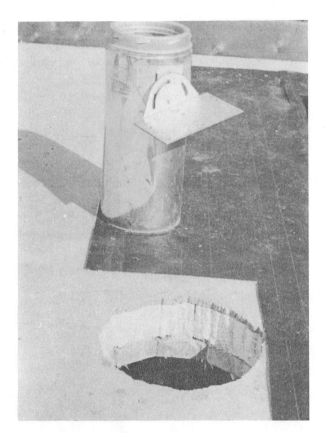

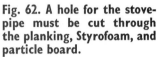

Fig. 62. A hole for the stove-pipe must be cut through the planking, Styrofoam, and particle board.

Design. The book recommends that all rigid insulation, such as Styrofoam and poly-urethane, be placed on the *outside* of the waterproof barrier, as we have done in the case of our walls:

> Insulation placed inside a waterproof or vaporproof layer will cause the surface temperature of that layer to cool substantially since it will be on the cold side of the insulation. Any water vapor permeating outwards through the insulation will then meet a cold impermeable layer which may cause condensation, an increase in the moisture content of the insulation, and hence a drop in the effective long-term R value.[6]

Well, I was a little worried when I read that in June of 1978. Had we made a big mistake? I ran right to the sauna, where I could still see the various roof layers exposed. The particle board and the Styrofoam were still in good condition, although I cannot say with certainty that there was no condensation at that point during the winter. If there was a moisture condition during the winter, it certainly was not evident and did not affect the planking. It is possible that water vapor did not penetrate our ship-lapped Styrofoam. Still, the advice quoted above should be carefully considered. Most, but not all, of the houses shown as examples in *Earth Sheltered Housing Design* do put the insulation outside of the membrane. Consider too that Styrofoam on the out-side will help protect the membrane from deterioration due to cold or earth movement. Some sort of hardboard should be used to protect the Styrofoam.

Waterproof Membrane

Knowing little about sod roofs, we followed fairly closely the advice given in an article, "The Return of the Sod Roof," by Hal. M. Landen (*The Mother Earth News*, No. 18). In the article, Landen advises covering the roof with 50-pound smooth-surfaced or mica-surfaced asphalt roll roofing:

The roll roofing is applied in horizontal strips, starting at the bottom of the slope and working up. The lower edge of each strip should overlap the sheet below by three inches, the lap itself should be sealed with salvage cement and galvanized roofing nails should be spaced every six inches along each lap and around the perimeter of the roof. Once you've worked your way—strip after overlapping strip—up each slope to the roof's peak, apply a generous layer of salvage cement to the entire surface. This sealer is also called *double coverage* or *black plastic* cement. Do *not* let a salesman sell you roofing tar.

Allow the cement to dry for a couple of hours. Its top will crust over but the underneath will remain soft and flexible (creating a resilient bed for the sod).

Next, spread a continuous sheet of black six-mil polyethylene plastic across the dried cement. The poly is available from any agricultural supply dealer and you *must* be sure to use black (which will last fifty years) instead of clear (which breaks down much sooner). You'll probably have to walk across the plastic as you smooth it down and—if you do—walk carefully to avoid putting holes in the sheet.

Although not absolutely necessary, it's a good idea to apply another coating of cement and a second layer of plastic over the first . . . just for insurance. There's no need to coat the top of the final poly sheet, however; grass roots will grow through the cement and the roll roofing, but they won't grow through the plastic.

We varied from the method quoted in only two ways. One, we stapled 15-pound felt instead of the 50-pound asphalt roll roofing. I figure that the real waterproof membrane is the plastic and that if water (and therefore grass roots) found its way to the asphalt, we would already be in trouble. I used the felt so that the plastic layers would not be applied directly to the hardboard, fearing that any expansion or contraction in the hardboard would exert forces that might rip the membrane. Now that I am familiar with the qualities of the plastics involved, I feel that the felt is not necessary. In fact, it gets in the way. We were constantly lifting felt inadvertently while trying to stretch the 6-mil plastic tightly. Therefore, either use the 50-pound roofing, as Landen advises, or leave this layer out altogether. The roll roofing probably improves the roof somewhat, but it will cost about $100, with nails. Nails, by the way, are normally spaced every 2 or 3 inches on roll roofing, not every 6.

The other slight deviation we made from Landen's recommendations was that we did not wait for the black plastic cement to dry before applying the 6-mil polyethylene. We would goop up a 3-foot strip the whole width of the roof, enabling us to reach over and grab the roll of plastic. This way, we could stretch the 6-mil without walking in the goop . . . which is like walking on greased ice (Figs. 63 and 64).

But now I must report that I'm not sure we took the easiest, best, or even cheapest route. Applying all that plastic roofing cement—we used 250 gallons—is a thankless and unbelievably messy job. It'll cost you a set of clothes and several pairs of gloves.

87

Figs. 63 and 64. Without waiting for the black plastic cement to dry, we applied the 6-mil polyethylene.

Unrolling and stretching a 32-foot-wide roll of plastic is no treat either. All in all, I wouldn't recommend this work to anyone I wanted to keep as a friend.

We applied a double layer of both plastic cement and polyethylene to the roof, a single layer of each as added protection for the walls. The Styrofoam, which should be applied just before the excavation is backfilled, is spot-glued to the wall with plastic cement.

If I were doing the roof again, I would try a different method, one used and recommended by John Barnard, the designer of Ecology House. Barnard applies three layers of 60-pound asbestos felt roofing, with hot pitch mopped between plies.[7] That too may sound like a lot of hot, messy work, but it's got to be easier than the black plastic method. In terms of price, there probably isn't much difference. Which is better? I don't know. Our roof doesn't leak. John Barnard's roof doesn't leak. Take your pick. Or research further. Some good literature has been published on all aspects of underground housing since we began Log-End Cave, and waterproofing is one area in particular where advancements seem to be coming quite quickly.

Earth Sheltered Housing Design suggests other waterproofing methods that are said to be superior to either the polyethylene layers or the built-up pitch membrane. Malcolm Wells, for example, has had success with $\frac{1}{16}$-inch butyl rubber sheets, completely bedded, but this is expensive. A very promising development is bentonite panels and spray-on bentonite. Bentonite is a clay which expands greatly when it comes in contact with water, which makes it self-sealing against further water penetration. It is a natural material and does not deteriorate with time. Bentonite is moderately expensive and not always easy to find away from large population centers. For further information on this and other waterproofings, see *Earth Sheltered Housing Design*.

Flashing

Flashing is one of the most important jobs in waterproofing and one of the most difficult to do well. It is necessary wherever there is a likelihood of water finding its way into the house or the framework, such as around skylights, stovepipes, retaining boards, and plates.

Keep a common-sense approach to the principle of flashing, which is essentially that of shingling: water runs down a slope. With a sod roof, though, another danger is present, which is that the soil may hold water against the flashing, allowing it to percolate *upwards* by capillary action. For this reason, it is necessary to apply a generous layer of plastic roofing cement between the flashing and the waterproof membrane. Figure 65 shows how we flashed the retaining boards by this method.

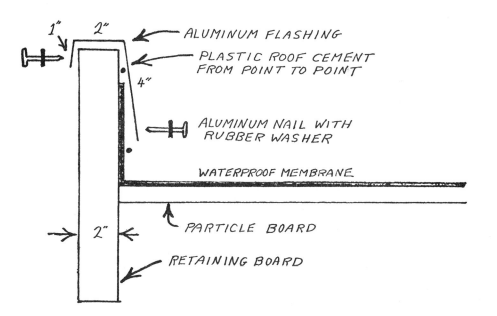

Fig. 65. The retaining boards were flashed by applying a generous layer of plastic roofing cement between the flashing and the waterproof membrane.

Flashing the plates on the north and south walls involves similar principles and methods. My method—and I don't know if it is necessarily the best—is to take the aluminum flashing into the 1-inch space where the thermopane panels will be placed. Later, the outside window molding will be nailed to the plate, compressing a bead of silicone caulking between the wood and the flashing. Where it is necessary to cut the aluminum flashing to fit around posts, be sure to use plenty of plastic cement or silicone caulking to seal the cut.

Flashing around skylights can be particularly bothersome, as the topmost surface of the "box" will be an attractive place for water to collect. Figure 66 shows the upper right hand surface of the boxed frame. The flashing should be of one-piece construction, and tucked under the waterproof membrane on the uphill side, but lapping over the membrane on the other three sides. The reader will note that we did not have a piece of flashing quite big enough to go around the box in one piece, so we had to lap another scrap on the downward side to complete the flashing. Always use plenty of overlap and roofing cement when piecing flashing together. The skylights can now be screwed into place according to the manufacturer's instructions.

For flashing the stovepipes, we used the tall flashing cone made by the manufacturers of our asbestos-lined stovepipes. Flashing is tricky enough, I feel, without attempting to build one's own piece to fit around a stovepipe.

Because of the overhang, the trickiest flashing of all is at the ends of the retaining boards, where the wood meets the soil. I advise the reader to leave this until last, so that he will have gained from his experience with the more straightforward flashing. The only additional advice I can offer concerning this particular problem—and I think it applies to all sorts of flashing situations—is to keep the lines of the structure beneath the flashing as simple as possible. Avoid projecting planks and compound angles, for example.

Fig. 66. Flashing around skylights should be of one-piece construction, tucked under the waterproof membrane on the uphill side and lapped over the membrane on the other three sides.

9. Backfilling

The key to a waterproof basement or underground house is good drainage. My own view is that good drainage is even more important than a good waterproof membrane. Give water an easier route to travel than through the walls and it will take that route.

Footing Drain

The key to good drainage is an effective footing drain, *and the key to a good footing drain is the use of a porous backfill.*

The footing drains should be constructed of 4-inch perforated plastic flexible tubing, preferably covered with a nylon or fiberglass "sock" to prevent the infiltration of sand. The tubing should be laid on a bed of washed 1-inch stone adjacent to the edge of the footing (Fig. 67). The drain should slope downwards slightly as it travels around the perimeter of the house and towards the soakaway. Our method of establishing this slope was to make marks with a crayon 4 inches below the top of the footing at the northwest corner, 5 inches below at the northeast and southwest corners, and 6 inches below at the low point, the southeast corner. We joined the marks by snapping a chalk-line and brought in washed stone, bucket by bucket, to the grade we had established. The slope, then, ran in both directions around the house from the high point at the northwest corner to the low point at the southeast corner, approximately 1 inch of drop for every 30 feet. From the southeast corner (Fig 68) the perforated pipe would be run at a similar slope behind the large retaining wall (discussed in Chapter 12) and on to a soakaway. (There is no reason why the retaining walls and drains could not be completed at this time, but we were in a hurry to get to the sod roof, still hoping to beat the winter weather.) The footing drains should be well covered with the crushed stone —say, 3 inches—and the crushed stone covered with 2 or 3 inches of hay (Fig. 69). The hay will decompose and form a mat to arrest the infiltration of the sand backfill into the crushed stone and footing drain.

The footing drains completed (for the time being), the only preparation before back-filling is to buttress the walls from the inside as a safety measure against the weight of the sand that will be pouring against the sidewalls (Fig. 70). This was easy, as we'd already framed most of the internal walls with 2-by-4 framing, and all we had to do

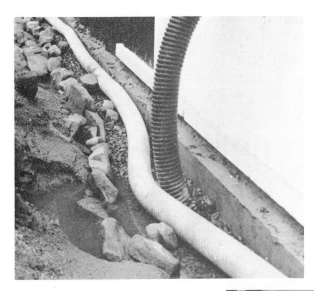

Fig. 67. The footing drain should be perforated, plastic, flexible tubing, laid adjacent to the footing on a bed of washed 1-inch stone.

Fig. 68. At the southeast corner—the lowest point of the footing drain's grade—the pipe will be run at a similar slope behind the retaining wall and onto a soakaway.

Fig. 69. Hay spread on top of the stones covering the footing drains will decompose and form a mat to arrest infiltration into the drain by the subsequent sand backfill.

was nail a diagonal brace to the framing for support. We did this at every point around the perimeter of the house where an internal wall met the block wall.

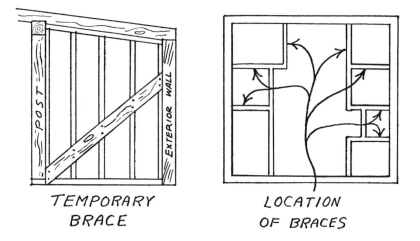

TEMPORARY BRACE

LOCATION OF BRACES

Fig. 70. Brace the frame from the inside to support it against the weight of the sand that will be poured against the sidewalls.

Backfilling

Because the material that came out of the excavation was relatively nonporous, it was necessary to bring in about 120 cubic yards of sand for backfilling. We probably saved $100 by using sand from our own pit on another part of the homestead. The sheets which the boys had cemented to the walls weeks earlier had come loose with the wind and the rain. I think it is best, therefore, to apply the Styrofoam immediately before backfilling. We used 2-inch Styrofoam for the first 4 feet of depth, where frost can be expected, and 1 inch below that to the footing.

The Styrofoam has two primary purposes. It keeps the wall warm, so that the concrete blocks act as a heat sink which helps to maintain a constant internal temperature; and it keeps the earth from freezing directly against the wall, which can crack a wall during a frost heave, and render almost any waterproofing ineffective. Of course, the sand backfilling should prevent such a heave, but I like to cover my bets.

We did not place the 1-inch Styrofoam on the wall where the food larder was to be, hoping that the uninsulated portion of the wall would encourage the transference of heat from the larder to the earth.

During the backfilling, be alert to the Styrofoam sheets moving or coming away from the wall, and watch for big stones that might come crashing down against the wall. Since the dozer driver may not be able to see such happenings from his perch on the machine, another pair of eyes is imperative.

Whenever insulation is to be applied to the exterior of a wall or roof in contact with the earth, always be sure to use genuine Styrofoam, also known as extruded polystyrene foam, which is highly resistant to moisture and frost deterioration. Beadboard, sometimes called polystyrene, and polyurethane foam are subject to deterioration (as well as loss of R-value) with unprotected applications.

10. Sod

This is the logical time in the construction of a subterranean home to put on the sod roof, as the winter will soon be approaching. An alternative job at this point is to spend a few days closing in the house against the elements: installing windows and temporarily shuttering the other rough openings with plastic or plywood, for example. We did this at Log-End Cave so that we could then fire up a working stove to help dry out the green roof rafters and timbers. The plantation pine we had to use when the hemlock ran out was especially troublesome; it was already starting to turn blue-gray with surface mold. We stopped its further deterioration well in time, however, and the end result is that the randomly-spaced pine boards offer a pleasing contrast to the predominantly hemlock ceiling. We were lucky.

There are two distinct approaches to sod roofing; planting the roof *in situ* or bringing in cut sods for placement on a bed of soil. Sowing seeds on the roof itself should be done in the spring. We chose the cut-sod method for two reasons. One, spring planting did not work in well with our plans; we wanted a thick green roof as early as possible in the summer. Two, I was concerned about erosion down the slope while the seeds were establishing their roots. I suppose there are ways to prevent this, such as the application of a thin mulch of straw or grass clippings. I must say that planting the roof *in situ* would probably have been less work, but I don't know if we would have obtained as good a roof as quickly, and we would have had to lay out more cash for topsoil.

To describe the placement of our sod, I must go back a few months to June. At the front of our homestead is a one-acre field which we had planted to buckwheat the year before to help rot out the "knotgrass" prevalent in grazing meadows in upstate New York. Although the year's growth of buckwheat had helped tremendously, the knotgrass, with its great intertwining root system, was already starting to form among the dried-up stalks of the buckwheat. I hired a roto-tiller and tilled a 30-by-70-foot corner of the field, cross-tilling in the opposite direction. A neighbor used the tiller on his garden while Jaki, a friend, and I removed stones and clumps of knotgrass from the plot for the rest of the day. As the tiller was on 24-hour hire, I was out first thing in the morning to cross-till the plot again. We spent a few more days in intensive removal of stones and knotgrass.

On June 12, friends helped us broadcast the timothy and rye seeds. We covered the seeds by gently brushing the ground with "brooms" made of young birch saplings which had been cleared from the site earlier. Later, we rolled the plot and applied a thin layer of mulch and wood ash. A week later the rye starting showing; after another week, the timothy. I had been advised to use timothy and rye by Tim Rice, a sod roof advocate from Vermont. The rye is for quick growth; the timothy will eventually take over as the predominant grass. The reader may wonder why we didn't simply use the knotgrass which was already in the field. The idea did occur to us, but we found the knotgrass impossible to cut and remove. The roots were too tough and too deep. Besides, we wanted to remove as many of the stones from the soil as possible, to lessen possible damage to the waterproofing.

I decided on a 6-inch sod roof, reasoning that any less would be especially susceptible to drought, while a thicker roof might exceed the design limits of the framework. I found that 2 to 3 inches of earth comes up with the cut sod, so I set out to lay a bed of 4 inches of good topsoil all over the roof.

We bought the topsoil from a local farmer. I was impressed with the rich, organic nature of the soil, and I thought it would drain well because of the fine sand mixed in with it. I was wrong, and should have guessed as much when I learned that the material had come from the bottom of a bog. Organic it was, but in dry weather, I'd mistaken silt for fine sand. It was a few weeks later, while repairing a flashing leak around one of the skylights, that we discovered just how poor the drainage really was. Oh well, I thought, at least the stuff will retain moisture during a drought. But that's a rationalization. I must advise the reader to use a soil with good drainage or, better, yet, to apply 2 or 3 inches of sand first, then a mat of hay, and then the topsoil. I would rather have a roof with good drainage, even though watering may be required more often.

We were careful to pass the first $\frac{3}{4}$-inch of topsoil through a $\frac{1}{2}$-inch screen so that no stones would lie directly on the waterpoof membrane (Fig. 71). After that, it was simply a matter of hauling load after load in the wheelbarrow and tamping frequently until we had a good 4 inches all over the roof (Fig. 72). Dennis, whose training is in golf-course maintenance, advised me that it was too late in the year (nearly the end of October) to put the sod on. I was inclined to agree, but wanted to do something to approximately the effect of a 6-inch sod cover in order to test the heating characteristics of the house in winter. I also wanted to know for sure what *would* have happened if I'd put the sod on at the end of October. On most of the roof, then, we spread a 2-inch layer of hay, not only to make up the insulation value of the missing sod, but also to prevent erosion of the 4 inches of topsoil during the spring thaw. As protection against the wind, we spread pine boughs over the hay and weighted the whole mess down with long sticks 2 to 3 inches in diameter.

As a test, I cut and laid about 20 square feet of sod near the peak of the roof. Dennis was right. The test patch was still in bad shape late the following June. At the same time, the rest of the roof, sodded in early June, was flourishing and in need of its second mowing.

Fig. 71. Sift the first $\frac{3}{4}$ inch of topsoil through a $\frac{1}{2}$-inch screen so that no stones lie directly on the waterproof membrane.

Fig. 72. A 4-inch layer of topsoil covers the roof.

Sodding was hard work, but not as bad a job as I'd expected. Three of us—one cutting, two hauling and laying—applied the sod in two days. We found that 10-inch square sods were the best size; larger squares tended to break up in transport. Sod should be cut when it is damp but not soaking, so that it will hold together and establish itself in its new location better. We hauled the sod with our pick-up. Laying the sod is easy. With the side of your foot, simply kick each new piece against those already laid (Fig. 73).

We accidently discovered what may be the easiest way of all to start a sod roof. The hay we'd used as insulating mulch was full of seeds and the roof was becoming quite green on its own in the spring. Unfortunately, we still needed another 2 or 3 inches of earth cover. Had we known, we might have applied the whole 6 inches of soil in the fall, and let the hay take over in the spring.

Fig. 73. Laying sod is easy: with the side of your foot, simply kick each new piece against those already laid.

11. Closing In

Closing in Log-End Cave required four different tasks: installing windows, building and hanging the door, installing vents, and infilling with stovewood (log-end) masonry.

There are seven windows in our house, not counting the skylights or the small thermopane in the door. Each of the seven is made of 1-inch thermopane—two panes of $\frac{1}{4}$-inch plate glass enclosing a $\frac{1}{2}$-inch dead air space. (Nitrogen is sometimes used in this space to aid moisture control.) All the windows were custom-made for the job by a local firm specializing in insulated windows.

For the windows to function at their full insulative value (and to allow for expansion), it is necessary to leave a $\frac{1}{4}$-inch dead-air space all around the unit. This means that the size of thermopane unit to order will be $\frac{1}{2}$-inch less than the rough opening, providing the opening is regular to begin with. Therefore, it is best to wait before ordering the units until the opening is made, so that the *actual*—rather than the *theoretical*—dimensions can be quoted.

Make an exception to this approach if you are able to score a bargain on thermopane units that were cut to the wrong size for another job. It's incredible how often this happens and, if the manufacturer is anything like our local fellow, he may be willing to sell such "mistakes" at greatly reduced prices. We double-glazed Log-End Cottage for $161 that way. If you use such units, make sure that the openings you frame are $\frac{1}{2}$-inch greater in each direction than the units you have found. It's worth the bother; new thermopanes are expensive.

I should caution against using single-pane windows at all. The condensation problem would be horrific and the heat loss entirely out of keeping with a home whose major practical advantage is its heating (and cooling) efficiency. Even our skylights are thermopane acrylic plastic. A double-glazed window, properly installed on the south wall of a house, actually admits more heat than it loses, even if shutters or insulated drapes are not used.[8] Unfortunately, even thermopanes are big heat losers when installed on the north side of a house. Luckily the snow drifts right over our north-facing windows, minimizing heat loss on that side during the winter.

We installed the three large rectilinear windows first. The procedure is (1) to install the external molding, (2) to place the window, and (3) to install the internal molding.

Installing the External Molding

We used rough-cut 1-by-1-inch stock for our window molding, both inside and out. Aesthetically, it is in keeping with the rough-hewn style of our house. And it's cheap. Ours was a by-product of our hemlock roof. Install the topmost piece first, figuring beforehand where you want the window in relation to the width of the surround. (Centered? Towards the outside for an internal shelf?) Cut the molding to length. Set your nails (I used 8-penny cup-headed) so that they are just breaking the surface on the side which is to be against the lintel. Run a bead of silicone caulking along the edge where the nails are breaking through. Nail the piece in place. The silicone will compress, giving a permanent seal against draft. Use *only* silicone caulking on work you want to last. Yes, it is expensive—$30 to $40 to caulk six to eight windows. But money spent on cheaper caulking is money poorly spent. And shop around. I've seen silicone vary by a dollar a tube in different stores in the same shopping center.

It is a good idea to leave the nail heads sticking out so that the piece can be removed for minor adjustment if necessary, at least until you are sure that the unit will fit flush against the molding. The most important consideration is laying the molding dead straight. I used a known straightedge to butt against for this purpose.

Now, using gloves and the aid of two helpers, put the unit in place against the top molding, make it plumb, and mark the sill with a pencil run along the outside edge of the glass. Put the unit back in a safe place (Fig. 74).

Fig. 74. Keep the window units on planks in a safe place while marking the sills and molding for installation.

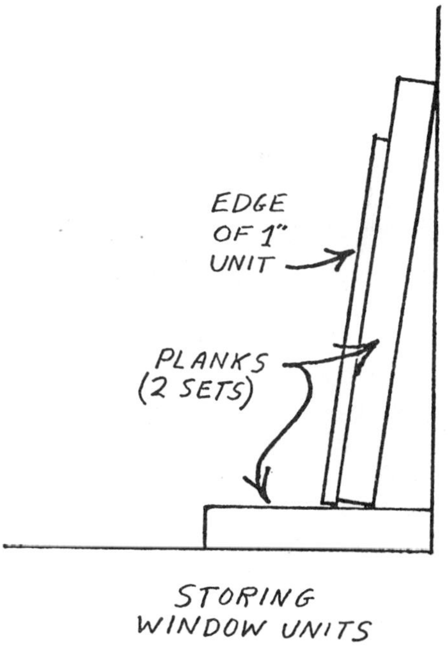

EDGE OF 1" UNIT

PLANKS (2 SETS)

STORING WINDOW UNITS

The bottom molding can now be installed, followed by the two connecting side molding pieces. Again, do not drive the nails home in case adjustment is necessary. Check again with the unit itself. If it fits flush against the molding, remove the unit and drive all the nails home. Run a bead of silicone around the perimeter of the molding so that the glass will compress it when the unit is placed permanently.

The key in placing the window unit is to set the thermopane on ¼-inch wooden shims, leaving a dead air space between the unit and the surrounding framework. Two or three shims will suffice for a 7-foot window. The shims should be about 2 inches long and just a shade under the thickness of the unit so that the plate glass itself will rest on the shim without the shim getting in the way of the inner molding. Carefully place the unit on the shims and press firmly against the silicone bead. Use small pieces of scrap molding at each end of the unit to hold the glass in place while the internal molding is applied.

Installing the Internal Molding

This procedure is pretty much the same as the one described above, except that the unit does not have to come up and down. Start with the top and bottom strips of molding. Again, use silicone against the glass and between the framework and the molding. The easiest way to do this double caulking is to run a bead along the glass itself and, after the nails have been started, another bead along the molding surface which is to be placed against the sill or lintel (Fig. 75). Nail the molding, being careful to avoid smashing the window as you drive the nails home. I usually hold a piece of cardboard or linoleum against the glass as protection while nailing the inside molding. Another trick is to angle the nails slightly towards the glass. This gives more room for nailing

Fig. 75. Double-caulk the windows by running a bead of silicone along the glass itself and another bead along the molding surface to be placed against the sill or lintel.

Fig. 76. Preparing to install the trapezoidal window in the north wall.

and helps compress the silicone bead against the glass. When the top and bottom molding strips have been nailed, the temporary stops can be removed and the side molding installed.

Figure 76 shows work being done on the installation of the trapezoidal windows in the north wall. The only difference is that we worked from the outside, which meant installing the inside molding first.

The Door

You can buy a prehung, insulated and weather-stripped door unit or make and hang the door yourself. The first alternative is much easier and, if care is taken in the selection of a quality unit, a weathertight door will result. I was sorely tempted to go the factory-made way and we even spent an afternoon looking at various doors. Jaki convinced me that I could make a door more in keeping with the rough-hewn atmosphere that we were after. (Easy for *her* to say.) I took a day and built the door. It is 4 inches thick. Heaven knows what it weighs. We had to hang the thing with two barn hinges and reinforce the connection between the post it was to hang from and the 4-by-10 lintel above with lag bolts and an iron L-bracket.

I will not go into detail about our own door construction, as people who make their own doors like to do it their own way, but I do include a cutaway view of the door which is fairly self-explanatory (Fig. 77).

Steve and Bruce came back up one weekend in November to see how things were going. We put them to work, of course, moving in an old kitchen stove we'd bought at an auction and installing the home-made door with heavy barndoor hinges.

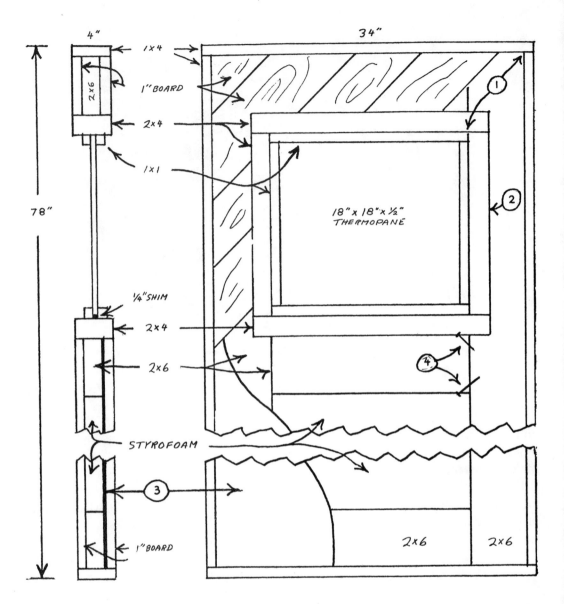

Fig. 77. Cutaway view of the hand-made door of Log-End Cave.

102

Installing the Vents

Ventilation was one of the unknowns we had to deal with at Log-End Cave. To what extent were vents necessary? Where should they be positioned? What size? How do you build them? Well, the subject wasn't covered at all in the few articles I was able to track down about underground housing, so I set out to design the most flexible system I could, with vents that I knew would be large enough to handle any situation short of a skunk in the stovepipe. The floor plan in Figure 78 shows the location of the six external vents: one in each bedroom, one in the sauna, one in the living room, and two in the kitchen-dining area. By opening and closing vents or doors, we can create practically any cross-draft situation we're ever likely to need for the removal of kitchen odors, for heating and cooling purposes, or simply to encourage a flow of fresh air.

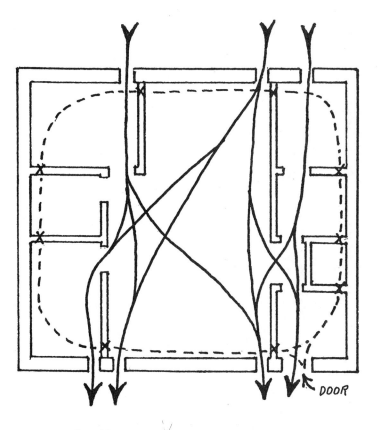

CROSS-VENTILATION PATTERNS

X = SUGGESTED LOCATION OF INTERIOR VENTS
(AT FLOOR LEVEL)

Fig. 78. Floor plan showing the location of six interior vents: one in each bedroom, one in the sauna, one in the living room, and two in the kitchen-dining area.

Our external vents are home-made. They are designed to fit unobtrusively into our 10-inch stovewood masonry walls. Construction is of aluminum screening and left-over 2-by-6s and 2-by-4s as shown in Figure 79. The vents are built on a flat surface, such as the concrete floor, creosoted on the outside surfaces, stained on the interior surfaces, and nailed into place. Jaki made the vent covers by gluing a piece of 2-inch Styrofoam to a piece of plywood corresponding to the shape of the vent surround. The Styrofoam is carefully shaped to fit the actual vent space. The cover can be held in place against the surround with a screw and a receiving socket. It can be decorated with material, paint, varnished wine-labels, whatever.

During the winter of 1978, we experienced no shortage of oxygen and did not need to use the vents at all, so we kept them closed to conserve fuel. The stoves must have been getting the oxygen they required from the understove vents. In spring and summer, we use the vents primarily to regulate humidity, opening them on dry days, closing them on humid days. (See Chapter 14 for a report on the need for floor-level interior vents.)

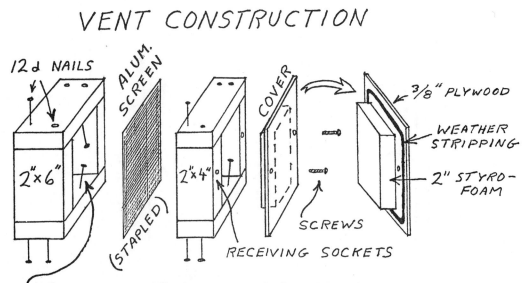

Fig. 79. Home-made vents can be constructed with aluminum screening, plywood, Styrofoam, weather-stripping, and leftover 2-by-4s and 2-by-6s.

Infilling with Stovewood Masonry

One of the design features of our house is the use of "stovewood masonry" infilling, which consists of short log-ends laid up like a stack of firewood and with mortar between the pieces. This construction technique, we feel, is conducive to the warm, rustic atmosphere we have tried to create. And it's about the least expensive way I can think of to close in a post and beam framework.

Our stonewood walls are of white cedar, 10 inches thick. While it is true that almost any wood can be used for log-ends, cedar is decidedly the best because of its high insulative value and its resistance to insects and rot. If care is taken to protect the wall —plenty of overhang is good policy—other wood can be used. If cedar is not available, use Douglas fir, western larch, or pine. Avoid white fir, spruce, or hardwoods, as they are more prone to rot. Consider also the insulative value of the wood you use. The insulative value is likely to be inversely proportional to the wood's value as firewood: that is, the heavier the dry wood, the greater the transfer of heat. The most important consideration in log-end building is to use completely dry wood—not almost dry, but *dry*. The log-ends should have quite a few cracks (or "checks") in them from drying. I'm always suspicious of any log-end which doesn't. The interior part of these checks can be stuffed with insulation to prevent drafts through the log-ends themselves. Air-drying of log-ends in the sun and wind is the best way, but this takes time, at least a year. The amount of log-ends required for our underground house was so little that I bought logs which had been cut three years before and "force-dried" them in our sauna at the cottage, and also at the cave itself, near the work stove.

The second most important consideration in log-end building is the mortar mix. I have experimented with three or four different mixes, but one (and its corollary) is clearly better than the rest: 6 sand, 8 sawdust, 2 Portland cement, 2 lime. A recipe of 6 sand, 8 sawdust, 3 masonry cement, and 1 lime amounts to about the same thing. These mixes are equally good, but of slightly different colors. The Portland mix is greenish-gray, while the masonry mix is bluish-gray. The key ingredients are the sawdust and the extra lime. The sawdust—which should be passed through a half-inch screen before mixing—retains moisture while the mortar cures, prolonging the drying. The longer it takes for mortar to cure, the less the likelihood of mortar shrinkage. Even using a sledge-hammer, I was unable to loosen a log-end from a panel in the sauna which had been laid up a few weeks before with the Portland mix described above. My friend Jack Henstridge, author of *Building the Cordwood Home*,[9] says that the extra lime calcifies and helps to fill in any cracks which might appear if log-ends shrink any further after being laid up. I use it because it gives a better mix for working with log-ends.

Cedar log-ends are themselves excellent insulation, but care must be taken to insulate the mortar, or the heat will transfer right out through the masonry. The sawdust in the mortar probably helps a little but, to be really sure of a warm wall, I recommend the fiberglass strip method of insulating described in the following pages. You can cut the fiberglass strips from a batt of insulation with a skill knife.

Figures 80 through 90 are taken from my previous book, *How to Build Log-End Houses,* which deals with the subject of stovewood masonry in great depth.

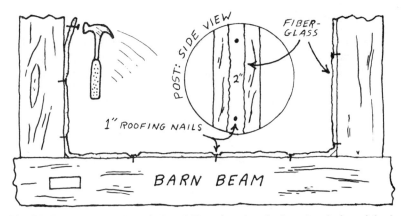

Fig. 80. Tack 2-inch-wide strips of fiberglass insulation that is $\frac{3}{8}$-to-$\frac{1}{2}$-inch thick along the plate beam and up the sides of the posts to prevent its falling into your work.

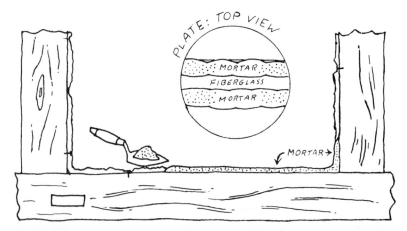

Fig. 81. Using a small trowel, lay a bead of mortar, $\frac{3}{8}$-inch-to-$\frac{1}{2}$-inch thick, along each side of the insulation on the plate beam and a few inches up the posts.

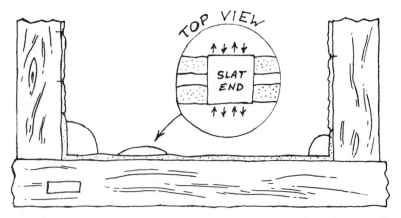

Fig. 82. Good pieces to start with are quarter log-ends and slat-ends. Lay them firmly in place, applying enough pressure so that the mortar squeezes out a little. A slight sliding motion back and forth at right angles to the plate will assure a good tight bond.

106

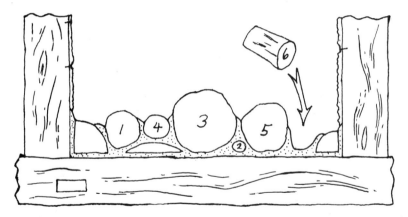

Fig. 83. Complete the first course, bedding and insulating up the sides of adjacent log-ends as necessary. Numbers indicate suggested order of placement.

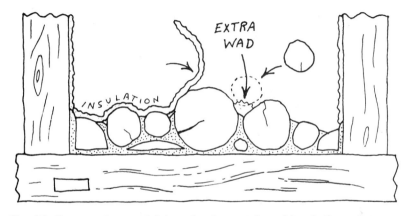

Fig. 84. To save time and labor, lay long strips of insulation whenever possible. Pack an extra wad in wherever a triangular space will be formed by a log-end capping two other log-ends.

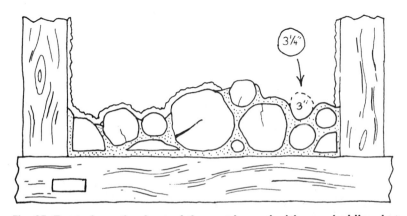

Fig. 85. Try to form the shape of the next log-end with your bedding, but just a tiny bit smaller than the piece you have in mind, so that it will force a little mortar out, assuring the best possible bond.

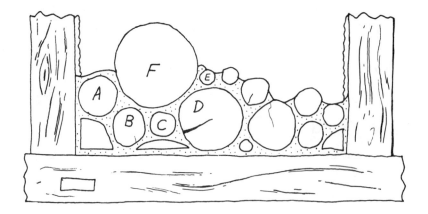

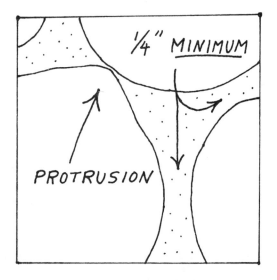

Fig. 86. Plan "cradles" for your big log-ends, again, so that the cradle is just a wee bit smaller than the end you are going to lay up. Log-ends A, B, C, D, and E, with the mortar bed above them, form the cradle for the large end F. Use a little pressure and that slight sliding motion to squeeze the mortar out a little. Don't be too neat with your mortar. Make a bit of a mess. A mess of mortar on the ground is a sign of a good bond. If you set the mortar board under your work and up against the plate, you should be able to salvage most of the mortar that falls off the wall. You can catch the stuff on the other side with your hand or the trowel. Wear gloves. Don't pressure your log-ends so much that the one you are laying up squishes down and touches the one below. Leave at least a $\frac{1}{4}$-inch of bed between ends where they are tangent; $\frac{1}{2}$ inch is better. A protrusion on a log-end can often be fitted into the triangular space enclosed by three ends, but sometimes there is no way to avoid a protrusion touching an adjacent piece and establishing the gap between them.

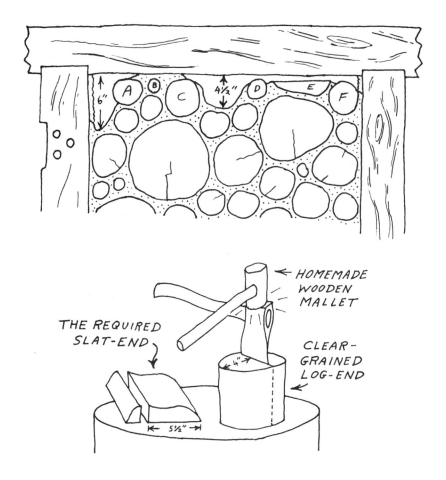

Fig. 87. It gets tougher near the top. A, B, C, D, E, and F are easy to find if you have kept a good selection of small ends and slat-ends handy. The upper left-hand corner wants a slat-end, but to get a good fit, you split a couple of inches off that is the right shape, but too big. The other space will be filled nicely with half a medium log-end. After bedding, take a measurement before you split, allowing $\frac{1}{2}$ inch for the topmost mortar joint.

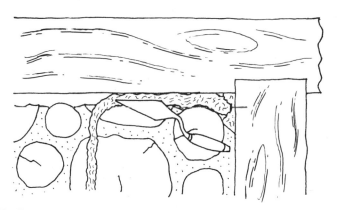

Fig. 88. Use the point of your trowel to stuff insulation over the last course.

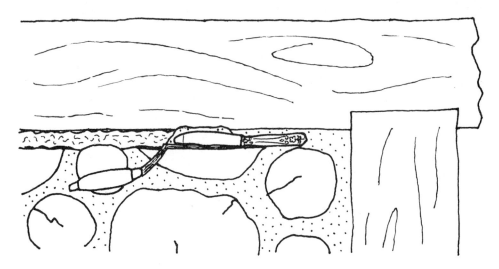

Fig. 89. To fill the final gap, push mortar off the back of the trowel with the pointing knife.

Fig. 90. The finished panel. The "beam-end" in the middle breaks the monotony of the round shapes. Pointing tidies up the panel and helps strengthen the mortar joints. When the house is finished, and the mortar is well seasoned, you can clean dirty log-ends and beams of any dusty mortar with a whisk broom. Particles of mortar do not adhere well to wood.

12. Retaining Walls and Landscaping

Because we were in a hurry to get the roof on, we backfilled before building the retaining walls. Actually, backfilling and building the retaining walls are very closely related: part of the function of the retaining wall is to keep the backfill from spilling around to the front of the house. In other words, we should have built the retaining walls first. As it was, part of the roof soil had to be retained with 2-by-6 planks on edge until the walls were complete.

The main cause of failure in retaining walls in northern climates is frost heaving. Therefore, good drainage behind the wall is imperative. For that reason, we continued the footing drains behind the large retaining wall and backfilled the wall with sand. I decided that the two small retaining walls would stand on their own without extra precautions since they were really little more than a single course of massive boulders. Even if frost was to shift them slightly, there's nothing to come toppling down.

We completed the drains before starting the actual stonework. The footing drains meet at the southeast corner of the house and continue as retaining wall drains until the large wall comes down to a single course of boulders. At this point the drain becomes nonperforated 4-inch flexible tubing, crosses the path leading to the front door, and continues on to the soakaway. The underfloor drains and the kitchen sink drain (graywater system) take a separate path to the same soakaway. The soakaway is a 10-foot diameter by 4-foot deep hole filled with fieldstones and covered with 15-pound roofing felt and earth (Fig. 91).

While discussing drainage, I must emphasize the importance of a good slope away from the front door. The frostwall footing should be exposed 2 inches above ground, and the pathway should fall away from the doorstep thus created by a good slope that drops an inch every 5 feet.

The retaining walls are built of massive stones that were removed from the excavation. This accomplished a dual purpose: an incredibly strong wall was built, and we got rid of a lot of pesky boulders that that would have made landscaping difficult.

As the average stone in the walls weighs several hundred pounds, there is little alternative to building the wall with a backhoe. There must be good coordination between the builder and the equipment operator on this sort of work. Before the operator arrives, spend an hour or so looking over the stones in the heaps. Then have the

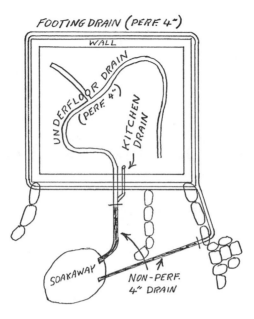

Fig. 91. The underfloor drain, kitchen drain, and footing drain all take separate paths to the same soakaway.

operator push all the good stones to a central depot near the work. This saves running around for the right stone when you need it.

The procedure is much the same as building any dry stone wall. Have the backhoe clear a flat base for the wall, perhaps depressed a couple of inches to give the large stones a bit of a bed in which to rest. Then, using a heavy chain with a ring at one end and a hook at the other, the backhoe can lift the required stone and set it into position. Use the stones with two parallel faces first for the early courses, saving stones with only one face for the top course. Shimming with smaller flat stones may be necessary during the actual building, so keep a pile handy.

After the backhoe is finished, the large triangular spaces that are bound to occur when building with such large stones can be filled and jammed with smaller stones (Fig. 92).

Not everyone's excavation will yield fine boulders to build with. It may be necessary to build retaining walls with smaller stones, blocks, bricks, or even old railway ties. The principles, however, are the same. Build on undisturbed earth or—better—a crushed stone foundation. Backfill with a porous material. Use perforated drain tile behind a high wall to carry away unwanted water.

The retaining walls complete, it is time to landscape. At our house there were two huge piles of earth to be spread out over the site, the largest on the east side. A bulldozer is easily the best machine for the job. Because we'd taken our easternmost retaining wall all the way to the old stone wall that surrounds the meadow, we had an ideal depository for excess earth. On the west side, there was just the right amount of earth to achieve natural contours away from the roof line. The desired effect is that the roof line looks as little like an unnatural protuberance as possible. This will not be one hundred percent possible with a sod roof only 6 inches thick. The peak formed by the intersecting roof faces is impossible to hide completely. But it is certainly possible for

112

Fig. 92. The large triangular spaces left between the boulders can be filled with smaller stones.

the roof to flow gently into the natural grade, and I think we have achieved this at Log-End Cave (Fig. 93).

Because of the lateness of the season, we left the spreading of the topsoil until spring. Then, after the sod roof was in place, we spread seven loads of topsoil to finish the landscaping. We planted timothy and rye and rolled the whole area.

The future will see rock gardens and shrubs as landscaping features. We have already grown snowdrops, an early spring flower, on the roof. Blueberry bushes set back 6 feet from the north gable end would be a practical way to decrease headlight glare from cars coming up the driveway.

Fig. 93. The finished product, seen from the windplant.

13. The Interior

There is nothing particularly special about building internal walls in an underground house. They are nonload-bearing and can be made of light materials. Simple stick framing of economy-grade 2-by-4s is perfectly adequate. Such a framework will take paneling, sheetrock, rough or finished boarding—almost any wall surface. We did most of the framing before the backfilling was done, so that we could buttress the walls as previously discussed. It is slightly easier to build the frame for a panel on the floor and then stand it up than it is to toenail each separate stud to the floor and ceiling plates (Fig. 94).

We made the floorplate fast to the floor with two or three 2-inch masonry nails. (Two-by-4s are commonly $1\frac{1}{2}$ inches thick.) Twelve-penny nails are fine for all other framing

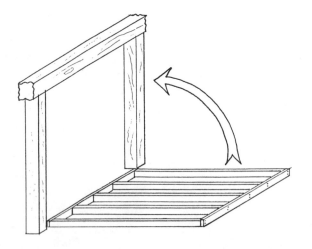

Fig. 94. When fashioning interior walls, it is easier
to build the frame for a panel on the floor and then
stand it up, rather than toenail each stud to the
floor and ceiling plates.

purposes. As discussed previously, our interior design was such that there was only one wall we had to fit around rafters. For the most part, we used ½-inch sheetrock (also called plasterboard) for the internal walls, painted with white textured paint. We make our own textured paint—a whole lot cheaper than buying it—by mixing one part cheap white latex paint with five parts premixed joint compound. Mix the ingredients thoroughly with a stick and apply with a roller. Scorching the roller first with a propane torch gives a nice three-dimensional look to the wall, but you'll use a lot more of the mix.

Our rafters are exposed and carried by three large barn beams, and the internal walls are planned to meet the underside of the beams. This leaves the problem of how to fill the space between rafters from the top of the beam to the roof planking so that sound will not travel from living to sleeping or bathroom areas. Luckily, there is a simple and practical solution, well in keeping with the motif of the house—log-ends. For these internal areas, we used 5-inch log-ends and laid them up with a full bed of sawdust mortar, no insulation (Fig. 95). Each panel has some different feature, such as symmetry, a specific pattern, or a particular kind of rubble effect. The use of stovewood masonry in this fashion ties the interior architecturally and aesthetically to the exterior and provides a particularly pleasing design feature to the main living-dining-kitchen area. And it's fun to do.

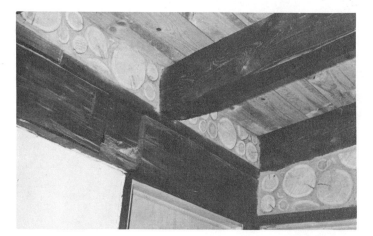

Fig. 95. Log-ends fill the space between rafters from the top of the beam to the roof planking and go well with the rough-hewn style of the house.

The Heat Sink

The room divider between the living and kitchen areas is of solid stone masonry a foot thick and, as both the kitchen and parlor stoves have their backs to the stonework, the room divider acts as a "heat sink," or "storage heater." A mass of stone 5 feet high, 8 feet long and, 1 foot thick has the capacity to absorb an awful lot of BTUs of heat, and because the stonework is at no point in contact with the exterior, the heat can only travel back into the room. The net effect is that the room divider has a moderating influence on the room temperature. Of course, the insulated block walls and even the floor are also heat sinks, but they operate at a lower temperature and take longer to "charge up."

Building the room divider was fun, I must admit, though I was hesitant about starting the project because I'd never built a free-standing stone wall before. Building the wall and the hearth took me five and a half days, with Jaki doing all the mortar pointing. We took the time to include several interesting design features, like stone shelves and a massive stone table (Fig. 96), stones with fossils in them, and a raised seat in the hearth in the shape of a butterfly (Fig. 97). My mortar mix was 5 sand, 1 Portland cement, 1 masonry cement. It is a good, strong mix, light in color, recommended to me by a neighbor who does a lot of stone masonry. The stones were taken from the huge stone wall in front of the house, which we had cut through with the bulldozer to provide vehicular access to the front door.

A detailed account of building a wall of stone masonry is beyond the scope and purpose of this book. If the reader wants to include a lot of stonework in his underground house, it would be worth buying or borrowing *Stone Masonry, The Owner-Builder's Guide* by Ken Kern and Steve Magers,[10] (The book is available at a cost of $6 from Mother's Bookshelf, P.O. Box 70, Hendersonville, N.C. 28739.)

Figs. 96 and 97. (Left) The room divider includes stone shelves and a massive stone table; (right) the hearth features a raised seat in the shape of a butterfly.

Wiring and Plumbing

These features are no different than for any other house. In our own house, the 12-volt wiring system connected to our windplant is particularly easy to construct. We do not have to bother with an earth wire, as the windplant itself is grounded and 12 volts do not give much of a shock anyway; and we do not have to meet the rigid inspection requirements of normal houses connected to Con Ed. Our plumbing system too, is a little different from most. It involves taking the water into the house with a push pump, filling a reservoir in the sauna, and using a 12-volt pump to carry the water, pressurized, to the bathroom and kitchen. As these are rather specialized systems which would not apply to the circumstances of most readers, I will not detail them here. Outside of siting considerations already discussed, no special conditions exist in an underground house that do not apply to any house or basement.

Floor Covering

Our floors are covered with linoleum in the kitchen and carpeting in all other rooms except the sauna and mud room, which are covered with a special paint for concrete floors. Because of our homesteading lifestyle and two German shepherds with perennially muddy paws, we chose a used industrial-grade (short nap) carpet in almost new condition. We carpeted half the floor area of the house for $120 plus underlayment. Friends in the trade laid the carpet and the linoleum in exchange for seven weeks free rent at the cave while we traveled to Scotland. A good deal all around.

Furnishing and decorating are matters of individual taste, of course, and I've really no intelligent comment to make in this regard except to advise keeping the walls and floors as bright as possible to maximize light.

14. Advantages

We have lived at Log-End Cave in both summer and winter. We have seen the outside temperature vary from −28° to 90° while the inside temperature during that time has never been below 58° or above 77°. The drainage and waterproofing systems met the severest possible test last spring when it rained steadily for three days on 3 feet of already compressed snow (Fig. 98). The only problem encountered during that time was one or two minor flashing leaks at the base of the front windows, easily repaired.

Other problems have occurred, but they were largely expected, and I will relate them in detail now.

Fig. 98. Winter at Log-End Cave.

"Energy Nosebleed"

A not quite expected problem, though I might have avoided it if I had read Malcolm Wells more carefully. On our south wall, at the corners, the concrete blocks are exposed on both sides of the wall. Because concrete is a terrible insulation, these blocks get very cold to the touch during the winter. The warm moist internal air hits the cold blocks and bingo, the dew point is met on the inside surface. Condensation is heavy. Cure: Styrofoam or wood on the outside allowed the blocks to store some internal heat and prevented a recurrence of the condensation.

Leaks in Skylights

Proper flashing of skylights is tricky business. Our primary mistake, as we learned, was to use soil with poor drainage near the skylights. Drainage is of paramount importance in taking the pressure off any waterproofing system. We cured the problem by digging up all the earth within a foot of the skylights and the main stovepipe. We then cleaned and dried the area thoroughly and applied plenty of roofing cement and 6-mil black polyethylene, well lapped. Finally, we used sand to backfill the skylights and spread wood chips over the sand for the sake of appearance (Fig. 99). The sandy area around

Fig. 99. Sand covered by woodchips provides the good drainage necessary to prevent leaks around skylights.

the large living-room skylight is connected to the bathroom skylight area by a 4-inch non-perforated flexible drain pipe imbedded in the intermediate topsoil. I carefully placed loose hay over both ends of this pipe to prevent infiltration of sand and soil. The bathroom and office skylight areas are similarly drained to the sand-backfilled area adjacent to the east and west walls. The sandy areas around the skylights, then, are finally drained to the primary footing drain system. The 4-inch pipes are sod-covered and invisible.

119

Pressure on Retaining Board

Our sod roof exerted a lot of lateral pressure on the 2-by-12 retaining board, probably because of expansion of the wet soil during the winter. By May, we could see that the board was almost an inch out of plumb. The cure was not difficult, and the reader can apply the fruits of our lesson right from the start. We removed the topsoil within 5 inches of the retaining boards, right down to the membrane. We replaced the soil with an inch of sand—to protect the 6-mil poly—and 5 or 6 inches of washed crushed stone. By this method, the 5 inches next to the retaining board are well drained, removing the hydrostatic pressure (Fig. 100).

Fig. 100. Removing the sod within 5 inches of the retaining board and replacing it with a layer of sand covered by 6 inches of crushed stone improved the drainage and relieved the lateral pressure from the retaining board.

Condensation

In late spring we began noticing condensation around the base of the wall. Here's what happens: Because of the so-called "flywheel effect," the earth's temperature is at its lowest in late May. Cold is conducted up through the footing to the area where the poured floor meets the first course of blocks. Warm, moist air hits the area—dew point, condensation, damp. Cure: Dehumidify the air. Allow good circulation of air around the perimeter of the house by the use of 6-by-8-inch vents at the corners where internal walls meet external walls. (See the cross-ventilation diagram, Fig 78.) If you have electricity, it is probably worth buying a dehumidifier for use during the first spring or two. Otherwise, fire up the woodstove every two or three days.

Two of the four problems discussed above involve warm, moist air. "Ah *ha!*" says the closet skeptic who has stayed with me this long. "I *knew* underground houses had to be damp!" Negative. But the concrete footings and floors (and the walls, in a poured-wall house) take about two years to cure completely. All the well-known underground architects report the same phenomenon. John Barnard says, "Don't get nervous. Whenever possible, leave the building open on dry days. Drying out may take two years."[11]

120

We didn't build with poured concrete walls, but we do have over 25 cubic yards of concrete in our structure, and our rafters and roof planks were very green when we put them up. All that moisture in the wood can only come into the house, because of the waterproofing. Wood takes about a year per inch to dry through side grain. We look for a much less humid condition next spring; the final humidity levels for 1980 should be about 50 percent in winter, perhaps 70 percent in summer on muggy days. We consider this to be healthy. And I must make it clear that the condensation problem we encountered during late May of 1978 was not really all that bad. Yes, we had to pull up the edges of the carpets, which were damp, and dry the place out. But within ten days we'd lowered the relative humidity from 90 percent to 75 percent by leaving the vents and doors open during the day and building an occasional fire. Below 80 percent, there is no problem, and once the hemlock roofing and rafters had a chance to dry out some, the humidity went way down.

The livability of Log-End Cave has exceeded our expectations. There is plenty of light; heating and cooling characteristics are phenomenal; the view to the south is excellent; one-floor living is both safer and more pleasant than stairs. There are unexpected joys, too. Like the sun's rays on summer evenings giving natural illumination to the dartboard. And being able to lie back in the bathtub bathed by sunshine from above. Of course, it is possible for folks to walk across your roof and peer into the skylights, but they are unlikely to do this more than once.

I said at the beginning of this book that heating and cooling characteristics and the low cost of construction are the primary advantages of subterranean construction. Let's examine the final returns on these matters.

Heating

Until January 20, 1978, we burned building scraps and a few logs to keep Log-End Cave at a working temperature of about 56°. On January 20, we brought a full measured cord of small-diameter hardwood into the mudroom and next to the stoves to dry. The outside temperature was 6° and the front door was opened a good part of the morning. The wood was ice cold from the woodshed and, stored in the mudroom, would take several days to reach room temperature. In the five hours from noon until 5.00 P.M. we brought the temperature from 60° (56° in the bathroom) to 70° (65° in the bathroom) by keeping both the cookstove and the parlor stove fired. The internal doors were not yet hung, so the transfer of heat around the house was very free. We had begun to charge the walls and the heat sink with BTUs. We were still living in the cottage, but we continued to keep the cave up to living temperature from January 20 onwards.

From February 2 to 6, we experienced constant subzero temperatures, day and night, bottoming out at −23° at 7:00 A.M. on February 5. During that time, I fired the parlor stove twice a day and the cookstove once. The temperature range during those five days was 58° (the cold morning) to 68° in the living room, 56° to 64° in the bathroom. The internal door from the mudroom was still not hung, and we later learned that that door made a big difference in heating the interior.

We moved into the cave on February 23. It was a sunny day and 23° outside. At 1:00 P.M., the inside temperature was 65° from passive solar heat alone. No stove had been fired since the previous day. We did not fire the cookstove until 2:30 P.M., when the sun had begun to move around to the west.

We found that on sunny days, cooking was all that was necessary to keep temperatures around 71°. We fired the parlor stove only on extremely cold nights. On March 6, we finished our first full cord of wood. It had taken twenty days to use the first half cord and twenty-six for the second half. Temperatures had remained fairly constant, both inside and outside, but the wood was stretching further because of charging the walls with heat, covering the floor with carpeting, and, finally, installing an internal door to the mudroom. That door had the effect of raising the interior temperature by an average of two degrees, while the mudroom, of course, went down to the fifties. By April 5, after the coldest February-March in the eleven years that I have records for, we had used 1.5 cords of wood. By June 1, the total fuel consumption was about 1.8 cords since January 20. According to the Miner Center in Chazy, New York (we live in West Chazy), the total degree days for the heating season from September 1, 1977, until June 1, 1978, were 8,645.* From January 20 to June 1, 4,527 degree days were recorded. Peter Gore, associate professor of Environmental Science at Plattsburgh and co-author of the study from which these figures were taken, says that it is a reasonable use of degree day figures to say that the amount of fuel required to heat for a part of the season is to the degree days of that part as the total fuel requirement for the heating season is to the degree days for that season.

In our case, then, we can set up the following relationship:

$$\frac{4{,}527 \text{ degree days (January 20–June 1)}}{8{,}645 \text{ degree days (September 1–June 1)}} = \frac{1.8 \text{ cords}}{X \text{ (total cords required for the heating season)}}$$

By transposition:

$$X = \frac{8{,}645}{4{,}527}(1.8 \text{ cords}) \qquad X = 3.43 \text{ cords}$$

But 1977–1978 was 3 percent colder than the average (1967–1977) degree day reading of 8,390 for a season, so we can adjust the figure to 3.33 cords of fuel required for the average winter. (Note: The four really cold months of December, January, February and March, which account for two-thirds of the total fuel requirement through June 1, were 6 percent colder than normal during the 1977–1978 season.) Also, because we were heating a bare house—no furniture, no floor covering—without internal doors for the first and coldest month of our test period, I think it is reasonable to assume a slightly better figure than this. I believe that we will stay warm over an average winter with three full cords of firewood or slightly less.

---

* The Heating Degree Day is a theoretical calculation based on the average number of degrees colder than 65° the temperature was at a given location over a 24-hour period. Sixty-five degrees is the accepted baseline for the Heating Degree Day calculation because it is assumed that if the temperature outside was 65° or warmer, no heat would be required from the heating system in the home. Residual heat, radiant heat from the sun, and cooking and other activities in the home would keep the temperature reasonably comfortable.

Since Heating Degree Days are calculated by determining how much colder than 65° it was on a given day, the colder it is the higher the number of Heating Degree Days will be. For instance, on a day when the average temperature is 40°, only 25 Heating Degree Days would be recorded. If the average temperature was 0°, however, 65 Heating Degree Days would be recorded. See Peter H. Gore and Judy G. Tedford, *Heating Degree Days: A Workbook for Homeowners* (Chazy, New York. Miner Center, 1978).

Even if one were to buy one's own wood at $40 per cord (we can get it in our area for $30), the total fuel bill would be only $120. Expressed in fuel oil at $.50 a gallon, this converts to $171. Or electricity at $.0325 per kilowatt-hour, $292.50. (See Appendix B for a comparison of fuel values.)

Why is it so easy to heat a subterranean house? What is its special magic? No magic, really—just common sense. Look at Figure 101. An above-ground structure is situated in ambient air, which in the winter commonly reaches −20° in our part of the world. If 70° is required in the house, it is necessary to heat to 90° above the temperature of the ambient. In an underground house, the ambient is the earth itself, about 40° in the winter where we live, which means it is necessary to exceed the temperature of the ambient by only 30°. The left side of the diagram in Figure 101 shows how a similar advantage is realized in summer cooling.

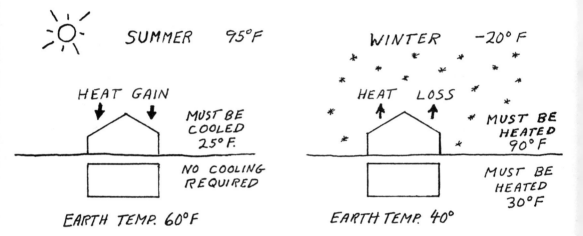

Fig. 101. The heating and cooling advantages of a subterranean house.

Cooling

There are three reasons for the ease of cooling a subterranean house in the summer. One is the "thermal flywheel effect." The earth's coolest temperature lags about two months behind the climate's lowest average temperature readings at a depth of 5.3 feet. The walls are likely to be coolest, then, in April, gradually warming up to the warmest earth temperature in September or October. (This also helps with heating during the first months of the heating season.) Of course, a negative factor of the cool April and May walls is the condensation problem discussed above, which is likely to occur for a year or two.

The second cooling advantage is that underground houses, with their characteristically slow temperature changes (thanks to the moderating effect of the mass of concrete and the surrounding cool earth), can effectively average out variations in temperature over a day, a week, or even a month. At a 5.3-foot depth in Minneapolis, earth temperature varies from 34° in April to 63° in September, so heat is conducted out of the house during all seasons.[12] In the summer, input from solar energy through sky-lights, excess lighting, people, dogs, cooking, and water heating are enough to offset heat losses through the walls, so that we are not too cool.

Finally, the earth and sod on the roof make a tremendous difference in cooling. When it rains, water is held by the sod and the soil, and it evaporates slowly. The evaporation has a cooling effect on the roof. In addition, sod is excellent for protecting the roof from extreme variations of temperature. The high surface temperatures recorded on roll roofing and asphalt shingles are unknown 6 inches below the surface.

To conclude: Subterranean houses do not require air conditioning, although some dehumidification may be necessary during the first two summers.

Subterranean houses use a lot less fuel than comparably-sized structures on the surface, though wood-burners should note that they will need a lot more kindling because (1) they won't be keeping their stoves on all the time, as they are inclined to do in surface houses, and (2) it will be necessary to build threshold fires much later into the spring than is necessary in surface housing. Of course, thanks to the flywheel effect, it will not be necessary to use the stoves nearly so frequently in the autumn.

Earth Sheltered Housing Design makes an interesting comment on the use of wood as fuel:

> Although wood is considered inconvenient as a primary fuel for a home with a large heating requirement, the low heating requirement of an earth sheltered home makes the use of wood feasible, especially if a method is provided for storage of the excess heat. A wood fired boiler coupled to a storage tank provides a heating system which can heat a home with only periodic fueling. If this system is used the stored heat must be delivered to the space to be heated through either radiation units or a forced air coil. This system can provide increased efficiency for the wood boiler as well as yielding a more even heat distribution to the space.[13]

Economy of Construction

Throughout the book I've been making reference to the low cost of underground housing. I realize that this contradicts a few of the underground architects and most of the evidence that has appeared in the popular periodicals of the last year or two, with the exception of Andy Davis's well-known 1,200-square-foot "Davis Cave," built in 1976. Interviewed in *The Mother Earth News*, No. 46, owner-builder Davis says, "The materials in the house itself cost us just $7,000. And the whole place—lots, house, carpeting, all new appliances, septic tank, everything—was only $15,000." And it should be pointed out that Davis Cave is made of poured reinforced concrete and has a rather elaborate electrical system.

The materials and contracting (excavation and landscaping) for Log-End Cave cost $6,750.57. Labor cost $660, bringing the total cost of the basic house to $7,410.57. Covering floors with good used carpets and new vinyl (concrete paint in the mudroom, sauna, and larder) added $309.89, fixtures and appliances another $507, bringing the total amount spent on our house to $8,227.46. I can assure the reader that this figure is within 3 percent of actual spending at Log-End Cave, as we have receipts for almost every item. An itemized accounting appears in Appendix A. The figure does not include the cost of our septic system, well, windplant, storage batteries, or furniture, all of which were already in use in the cottage. The cost of the septic tank and drain field in 1975 was approximately $600. The windplant, tower, and batteries cost about $800. The spring-fed well was already on the property.

Of course, these figures do not place a dollar-value on the owner-builder's own labor, but I believe that this should be measured in terms of time, not money. Money saved is a lot more valuable than money earned because we have to earn so darned much of it to save so precious little.

I figure that roughly 1,800 hours of work were required to build Log-End Cave. This includes about 450 hours of outside help, some of it paid, some of it volunteered. The bulk of the work was done during a five-month period ending on December 17. During January and February, we completed most of the interior work. Final landscaping and placement of the sod was done in June. For about half of the days, I noted actual hours worked in my diary, especially when this was required to pay help, but for the other half I have had to read my notes to see what was done and guess at the labor time. Therefore, I cannot be sure that the 1,800-hour figure is as accurate as the cost analysis, but I feel sure that it is a figure within 200 hours of the reality. Everyone works at different rates anyway, so the actual time it takes someone else may vary widely from the 1,800-hour figure.

One-third of the "average" American's after-tax income is devoted to shelter, usually rent or mortgage payments. If a man works from the age of 20 to the age of 65, it could be legitimately argued that he has put in fifteen years just to keep a roof over his head. With six months' work (and $6,750), he and his family could have built their own house.

"All that is fine and dandy for writers, golf pros, and other layabouts," people tell me. "But I'm a *working* man. I can't afford to take six months off to build my own house."

I reply that to save fourteen and a half years of work, they can't afford *not* to build, even if it means losing a job. Granted, the $6,750 has to come from somewhere, but this is no more—and probably less—than the down payment on a contractor-built home, and less than half the cost of a new mobile home of similar living space. And you can't compare the end product.

What do you get for your time and money? You get a comfortable, long-lasting, energy-efficient, low-maintenance dwelling. You get the design features that suit you, so that your house fits your personality like an old slipper. You get built-in fire, earth-quake, and tornado insurance. You get freedom from a life of mortgage payments. You get tremendous personal satisfaction. You get in on the *below*-ground floor of "an idea whose time has come."

APPENDIX

A. Log End Cave—Cost Analysis

Heavy equipment contracting	$892.00
Concrete	873.68
Surface bonding	349.32
Concrete blocks	514.26
Cement	47.79
Hemlock	345.00
Milling and planing	240.26
Barn beams	123.00
Other wood	167.56
Sheetrock	72.00
Particle board	182.20
Nails	62.88
Sand and crushed stone	148.21
Topsoil	295.00
Hay, grass seed, fertilizer	43.50
Plumbing parts	124.95
Various drain pipes	166.23
Water pipe	61.59
Metalbestos stovepipe	184.45
Styrofoam insulation	254.93
Roofing cement	293.83
Six mil black polyethylene	64.20
Flashing	27.56
Skylights	361.13
Thermopane windows	322.50
Interior doors and hardware	162.80
Tools, tool repair, tool rental	159.43
Miscellaneous	210.31
	————
Materials and contracting cost of house, landscaping and drainage	$6,750.57
Labor	660.00
	————
Cost of basic house	$7,410.57
Floor covering (carpets, vinyl, etc.)	309.89
Fixtures and appliances	507.00
	————
Total spending at Log End Cave	$8,227.46

B. Amount of Other Fuels Equivalent to a Cord of Air-Dry Wood

A Cord of Air-Dry Wood equals	Tons of Coal	Gallons of Fuel Oil	Therms of Natural Gas	Kilowatt Hours of Electricity
Hickory, Hop Hornbeam (Ironwood), Black Locust, White Oak, Apple =	0.9	146	174	3800
Beech, Sugar Maple, Red Oak, Yellow Birch, White Ash =	0.8	133	160	3500
*Gray and Paper Birch, Black Walnut, Black Cherry, Red Maple, Tamarack (Larch), Pitch Pine =	0.7	114	136	3000
American Elm, Black and Green Ash, Sweet Gum, Silver and Bigleaf Maple, Red Cedar, Red Pine =	0.6	103	123	2700
Poplar, Cottonwood, Black Willow, Aspen, Butternut, Hemlock, Spruce =	0.5	86	102	2200
Basswood, White Pine, Balsam Fir, White Cedar =	0.4	73	87	1900

Assumptions—

Wood: 1 cord = 128 cubic feet wood and air or 80 cubic feet of solid wood at 20% moisture content. Net or low heating value of one pound of dry wood is 7,950 Btu. Efficiency of the burning unit is 50%.

Coal: Heating value is 12,500 Btu per pound. Efficiency of the burning unit is 60%.

Fuel Oil: Heating value is 138,000 Btu per gallon burned at an efficiency of 65%.

Natural gas: One therm = 100,000 Btu = 100 cu. ft. Efficiency of burning is 75%.

Electricity: One KWH = 3,412 Btu. Efficiency is 100%.

*This is the quality of wood our tests are based on.

Source Notes

[1] Rudofsky, Bernard, *Architecture Without Architects* (New York: The Museum of Modern Art, 1964).

[2] Dempewolff, Richard F., "Your Next House Could Have a Grass Roof," *Popular Mechanics,* (March 1977), p.140.

[3] Dempewolff, Richard F., "Underground Housing," *Science Digest,* (November 1975), p.140.

[4] Labs, Kenneth, "The Use of Earth-Covered Buildings through History," *Alternatives in Energy Conservation,* p. 16.
An excellent, all-inclusive discussion of the subterranean idea, *Alternatives in Energy Conservation* presents the proceedings of a conference held in Fort Worth, July 1975. It is for sale by the Superintendent of Documents, U.S. Government Printing Office, Washington, D.C. 20402, at a cost of $3.25. Ask for stock no. 038-000-00286-4.

[5] Meyer, Kirby T., "Utilities for Underground Structure," *Alternatives in Energy Conservation: The Use of Earth Covered Buildings,* pp. 166 and 167.

[6] *Earth Sheltered Housing Design,* American Underground-Space Association, Department of Civil and Mineral Engineering, University of Minnesota, p. 147.

[7] Smay, V. Elaine, "Underground Living," *Popular Science,* June 1974.

[8] *Earth Sheltered Housing Design,* op. cit., pp. 71–73.

[9] Copyright and published by J. R. B. Henstridge, 1977.

[10] Charles Scribner's Sons, N.Y., 1977.

[11] Dempewolff, Richard F. "Your Next House Could Have a Grass Roof," p. 144.

[12] Bligh, Thomas, "A Comparison of Energy Consumption in Earth Covered vs. Non Earth Covered Buildings," *Alternatives in Energy Conservation,* op. cit., p.92.

[13] *Earth Sheltered Housing Design,* op. cit., p.68.

INDEX